FROM PHYSICS TO BIOLOGY

b i f i 2 0 0 6

II INTERNATIONAL CONGRESS

From Physics to Biology:

The Interface between Experiment and Computation

ZARAGOZA, SPAIN
FEBRUARY 8-II, 2006.

INVITED SPEAKERS

- M. Amzel (Johns Hopkins, *USA*)
- C. Cavasotto (Molsoft LLC, *USA*)
- S. Cocco (Ecole Normale Supérieure, *France*)
- E. Freire (Johns Hopkins, *USA* and BIFI)
- M. Karplus (Strasbourg and Harvard, *USA*)
- S. Leibler (Rockefeller, *USA*)
- A. Perczel (Eotvos, *Hungary*)
- A. Tramontano (La Sapienza, *Italy*)
- G. Waksman (Institute of Structural Molecular Biology, *UK*)
- M. E. Wall (Los Alamos National Laboratory, *USA*)
- E. Westhof (Strasbourg, *France*)

Organizing Committee

J.L. Alonso (Zaragoza), L. Arrachea (Zaragoza), P. Bruscolini (Zaragoza), J. Clemente-Gallardo (Zaragoza), A. Cruz (Zaragoza), F. Falo (Zaragoza), E. Freire (Johns Hopkins), E. Marinari (Roma), Y. Moreno (Zaragoza), J.M. Sánchez-Ruiz (Granada), J. F. Sáenz (Zaragoza), A. Velázquez-Campoy (Zaragoza), I.Vidal (Conference Secretary)

UNIVERSIDAD DE ZARAGOZA

INSTITUTE FOR BIOCOMPUTATION AND PHYSICS OF COMPLEX SYSTEMS.

The Organizing Committee promotes a call for contributions. There will be 11 invited lectures and around 16 short oral contributions to be selected among those presented as a poster when they are best suited for a talk. Please see the web page of the conference for instructions and further information.

http://bifi.unizar.es

iberCaja
Obra Social y Cultural

COMPUTER CLUSTER

IMPORTANT DATES

Deadlines:
Early registration:
October 15, 2005

Last day to complete registration:
November 15, 2005

Poster abstract submission:
January 5, 2006.

MORE INFORMATION:
http://bifi.unizar.es

Dates and venue:
February 8 - 11 2006
Auditorium of
Zaragoza

GOBIERNO DE ARAGON
Departamento de Ciencia, Tecnología y Universidad

MINISTERIO DE EDUCACION Y CIENCIA

FROM PHYSICS TO BIOLOGY

The Interface between Experiment and Computation

BIFI 2006 II International Congress

Zaragoza, Spain   8 – 11 February 2006

EDITORS

Jesús Clemente-Gallardo
Yamir Moreno
José Félix Sáenz Lorenzo
Adrián Velázquez-Campoy

*Institute of Biocomputation and Physics of Complex Systems,
University of Zaragoza, Spain*

SPONSORING ORGANIZATIONS
University of Zaragoza
Government of Aragón - (Diputación General de Aragón)
Ibercaja
Spanish Ministry of Science and Education
Megware Computer GmbH

Melville, New York, 2006
AIP CONFERENCE PROCEEDINGS ■ **VOLUME 851**

Editors:

Jesús Clemente-Gallardo
Yamir Moreno
José Félix Sáenz Lorenzo
Adrián Velázquez-Campoy

Institute of Biocomputation and Physics of Complex Systems
University of Zaragoza
Corona de Aragon 42
50009 Zaragoza
Spain

E-mail: jcg@unizar.es
      yamir@unizar.es
      jfsaenz@unizar.es
      adrianvc@unizar.es

L.C. Catalog Card No. 2006930967
ISBN 0-7354-0350-3
ISSN 0094-243X

Printed in the United States of America

CONTENTS

Foreword

Just a few lines as Director of the BIFI to thank all the participants of the Second International Conference BIFI2006 "From Physics to Biology: the interface between experiment and computation".

One of the main aims of our Institute is the interdisciplinary collaboration of chemists, physicists and biologists, of theoreticians and experimentalists. With this goal in mind we organize an International Conference every two years, in this occasion on the interface of Physics and Biology.

The Conference of this year has accomplished most of its goals, first with the excellent interventions of the Invited speakers, whom I would like to thank for their help and for having made easier our task as organizers. I would like also to thanks the rest of participants for their great presentations and the enlightening debates which followed both the oral communications and the poster sessions. All this together ensured the scientific success of the Conference. Let this volume be a humble tribute to all the participants, for their interest and their search for knowledge.

Finally I also want to thank the members of the Institute who generously offered their efforts and time to organize this event: theirs is a major part of the success of the Conference.

José Felix Sáenz Lorenzo
Director of BIFI

Preface

During the last several years, molecular and cell biology have attracted the attention of both biologists and physicists and they have moved from the study of individual components toward the modeling and understanding of many interacting components. On the other hand, the complexity of biomolecules and their collective behavior make necessary the interaction between those researchers devoted to experimental and theoretical studies. The aim of the BIFI2006 Conference was therefore to bring together researchers working at the interface between Physics and Biology from both the theoretical and experimental point of views. It was held by the Institute of Biocomputation and Complex Systems Physics (BIFI), at the University of Zaragoza, Spain. There were three main topics at the conference: Nucleic Acids, Proteins and Peptides, and Collective Behavior of Biomolecules. Specifically, the subjects covered included:

- Mechanical properties of DNA and RNA: unzipping and stretching of single molecules.
- Structure and folding of RNA.
- Non canonical forms of DNA.
- Docking of virtual ligand libraries containing millions of molecules.
- Conformational flexibility and ligand docking.
- Energy landscapes and ligand binding.
- Protein-protein interactions: the interactome.
- Signal networks and disease.
- Macromolecule folding with emphasis in structure prediction.
- Biomolecular interactions and their analysis with theoretical/computational approaches of relevance for ligand and drug design.
- Studies into the collective behavior of interacting biomolecules and the modeling of biological networks.
- Self -assembly of DNA structures and proteins.
- Cell Signaling.
- Ab initio and DFT molecular computations.
- Disordered systems and molecular and neural networks.
- Solvation energy.
- Energy functions.

The Conference attracted more than 120 scientists coming from many different countries and ran over 4 days. The list of invited speakers included leading scientists from all over the world: M. Amzel (Johns Hopkins University), C. Cavasotto (Molsoft LLC), S. Cocco (Ecole Normale Supérieure, Paris), E. Freire (Johns Hopkins University and BIFI), M. Karplus (Strasbourg and Harvard Universities), S. Leibler (Rockefeller University), A. Perczel (Eötvos University), A. Tramontano (University of Rome), G. Waksman (Institute of Structural Molecular Biology, UK), M.E. Wall (Los Alamos National Laboratory), and E. Westhof (Strasbourg University). They all gave lectures where the main directions of the aforementioned subjects were reviewed as well as outlined current trends in biological experimentation and theoretical and numerical modeling.

The present issue of the AIP Conference Proceedings aims to provide a glimpse of what kind of works were discussed during the Conference.

Cavelier & Amzel describe a computational study, using a combination of Quantum Mechanics and Molecular Mechanics, intended to elucidate the electron transfer mechanism catalyzed by the redox enzyme quinone reductase, emphasizing the critical role of dynamics and fluctuations. Wall contributes with a manuscript that is at the forefront of current theoretical research in the subjects of ligand binding, protein fluctuations and allosteric free energy. Cavasotto contributes with a very well-taken manuscript where the current knowledge and the last developments in virtual screening are thoroughly explained, pointing out also the strong and weak points. Cocco & Monasson present the modeling and characterization of mechanical unzipping of single nucleic acid and the effect of the nucleotide sequence; by simulating the experimental output, they explain how to approach the inverse problem: how to predict the sequence from the monitored signal. Giorgetti et al. review some of the current challenges in protein structure prediction.

Junier et al. shortly review the protein folding problem in small proteins in a work that may represent a good introduction to the beginners in the field and goes over different techniques, both experimental and theoretical, with emphasis in single molecule experiments; state of the art of models (at different length scales) as well as numerical work (mostly simulations) done on them are also addressed. Tortosa & Jaramillo have made an effort to communicate in a short contribution several technical aspects on computational protein design, in particular, the design of active sites into protein scaffolds which can be validated experimentally. The benefits from the interaction of biologists and physicists is illustrated in the work presented by Danenberg, who studies different energetic strengths and cooperativity of H-bonds in peptides by means of a technique borrowed from the physical sciences, namely, Density Functional Theory. Along the same lines, Echenique et al. analyze the effects of constraints in the conformational equilibrium distribution of peptides through quantum mechanical calculations, the definition of an appropriate set of internal coordinates for branched molecules, and a statistical measure of differences between different levels of potential energy approximations. Zaman performs a multiscale modeling of cell migration landscapes at various length and time scales. Sanchez-Ruiz presents a summary of some new visions on protein folding thermodynamics constructed from experimental evidences of downhill protein folding, focusing on deviations from the classical two-state behavior that are not due to intermediates, but to very low, or even absent barriers.

Hernandez et al. present a summary of recent findings on a very interesting and appealing issue: the complex regulatory mechanism of a protein, FurA, involved in iron regulation; the experimental methodology and tools, as well as open questions, are pointed out. A simple elastic model, the Gaussian Network Model, is used as a tool to study the differential looseness of different regions of the apoflavodoxin protein by

Bruscolini et al.; the model, upon introduction of the possibility of breaking contacts, is capable of reproducing the thermodynamics of the unfolding process, in particular, the presence of an intermediate during thermal unfolding. Moreno et al. review several topological and dynamical aspects of recent studies concerning the highly topical subject of biological networks at both molecular and cellular levels. Velázquez-Campoy describes the implementation of the McGhee-von Hippel theory for ligand binding to linear macromolecules with overlapping binding sites in isothermal titration calorimetry. Gracia et al. pay attention to ligand docking with consideration of the target flexibility at the side-chain and backbone level making use of NMR solution structures that would represent the different conformations adopted by the target. Schubert et al. report in a detailed and clear contribution a phylogenetic analysis of FAD synthetase aimed to identify key residues essential for substrate binding and catalysis. Samsonov et al. reports on the expression of the Ctr1 gene in specific tissues and propose an in silico model of the protein produced by this gene. Basdevant et al. deal with a model of solvation at an intermediate level between implicit and explicit methods. Finally, the characterization of the interaction between protein and metal ions, employing thermodynamic and structural tools and techniques, and the physiologic role of metal in regulating the function of the DtxR repressor are addressed by D'Aquino and Ringe.

ACKNOWLEDGMENTS

We would like to acknowledge all the sponsors of the Conference, and in particular, to *Universidad de Zaragoza, Diputación General de Aragón, IberCaja, Ministerio de Educación y Ciencia*, and *Megware Computer GmbH*. Heartfelt thanks also go to the Organizing Committee (specially to José L. Alonso), to all invited speakers that generously accepted our invitation, to the participants for sharing their results and for creating a scientifically rich environment and to the conference secretaries (Conchita Carbó, Mercedes Fatás and Isabel Vidal) whose hard work and efforts made possible this successful meeting.

Jesús Clemente-Gallardo
Yamir Moreno
José Félix Sáenz-Lorenzo
Adrián Velázquez-Campoy

Institute of Biocomputation and Complex Systems Physics (BIFI)
Universidad de Zaragoza
Corona de Aragón 42
E-50009 Zaragoza, Spain

Role of Fluctuations in Quinone Reductase Hydride Transfer: a Combined Quantum Mechanics and Molecular Dynamics Study

By German Cavelier[*][†] and L. Mario Amzel[*][¶]

[*] Department of Biophysics and Biophysical Chemistry
Johns Hopkins University School of Medicine, Baltimore, MD 21205
[†] Present address: Office of Interdisciplinary Research and Scientific Technology,
Division of Neuroscience and Basic Behavioral Science,
National Institute of Mental Health, National Institute of Health, Bethesda, MD 20892
[¶] Corresponding Author. E-mail: mario@neruda.med.jhmi.edu

Abstract: Quinone Reductase is a cytosolic FAD-containing enzyme that carries out the obligatory two-electron reduction of quinones to hydroquinones. The first step in the mechanism consists of the reduction of the FAD by NAD(P)H via direct a hydride transfer. Combined QM/MM calculations show that the protein accelerates this step by a combination of effects that include charge stabilization and distortion. The calculations also show that dynamic effects play an important role in QR catalysis: the distance between the donor and the acceptor atoms of the hydride transfer, which is too long for transfer in the static structure, becomes shorter than 3 Å 25% of the time due to motions of the protein and the cofactors.

Keywords: Density Functional, Enzymes, Flavoproteins, DT-diaphorase, isoalloxazine, Quantum Chemistry, Protein fluctuations, Molecular Dynamics

INTRODUCTION

Quinone reductase (QR1; NQO1), a cytosolic phase 2 detoxification flavoenzyme, catalyzes the obligatory two-electron reduction of quinones to hydroquinones using either NADH or NADPH as the electron donor. The catalytic cycle of QR1 involves reduction of bound FAD by NAD(P)H *via* a hydride transfer from C4 of the nicotinamide to N5 of the flavin. In a previous paper[1] the energetics of charge stabilization by QR1 following this hydride transfer was estimated using an *ab initio* DFT calculation. However, other important aspects of the mechanism of H⁻ transfer remain poorly understood. For example, in the x-ray structure of the complex of NADP⁺ with oxidized QR1, the distance between C4 of the nicotinamide and N5 of the flavin (4.2 Å), is probably too long for a direct hydride transfer. Since the hydride transfer is fast (i.e. is not rate

CP851, *From Physics to Biology; BIFI 2006 II International Congress,*
edited by J. Clemente-Gallardo, Y. Moreno, J. F. Sáenz Lorenzo, and A. Velázquez-Campoy
© 2006 American Institute of Physics 0-7354-0350-3/06/$23.00

limiting in QR1) questions remain about what effects contribute to the acceleration of this step.

It is becoming increasingly clear that enzymes are not rigid "lock and key" devices: molecular motions may make critical contributions to rate acceleration in enzymatic catalysis[2]. Enzymes can bring their substrates closer to the transition state geometry not only by static stabilization, but also by means of conformational fluctuations, dynamic preorganization, and other dynamical effects. Here we analyze the energetics of these effects in the hydride transfer step of QR1 by *ab initio* quantum mechanical calculations combined with molecular mechanics/dynamics (QM/MM).

METHODS

Computer Programs. Cofactors within the context of the protein were modeled using the programs QUANTA and UNICHEM on SGI workstations. The initial optimizations and transition state search were carried out with the MNDO 94 program (AM1 Hamiltonian), accessed through UNICHEM, and run on an SV1 CRAY at the Advanced Biomedical Computing Center, National Cancer Institute, Frederick, MD. *Ab initio* calculations were done as single point energy calculations at the B3LYP/6-31G(d) level of theory, as implemented in GAUSSIAN[3] 98 on the same SV1 CRAY supercomputer. Visualizations were done with UNICHEM, QUANTA, MATLAB, and EXCEL. Molecular Dynamics calculations were performed with CHARMM on a Power Challenger SGI.

Residues included in the quantum mechanical calculations. In addition to the cofactors, the following residues, determined to be important in the catalytic mechanism of QR1, were included in the QM calculations: Trp 105, Phe 106, Gly 149, Gly 150, Tyr 155, and His 161 [4-7]. All residues and residue pairs were terminated at their carboxy termini with methyl groups.

Identification of the transition state. The system considered here comprises approximately 200 atoms from FAD, NAD, and the interacting amino acid residues. It is not practical to perform a saddle point optimization for a system of this size[8-19]. Furthermore, as the reaction coordinate comprises movements of cofactors and of residues around the active site of the enzyme, finding the transition state requires optimization combined with molecular dynamics of the whole protein, i.e., combined QM/MM methodology[20-33]. To overcome these computational difficulties, we carried out our calculations in two stages: (1) identification of an approximate transition state, and (2) optimization of the system by successive approximations.

Starting with the optimized structure a series of grid calculations was made in points around the straight line from C4N to the transition state, and around the straight line from the transition state to N5F. Optimization at each grid point was performed allowing the cofactors to relax, with the exception of their links to the sugar-adenine portion, which were kept fixed at the x-ray coordinates. (This restriction mimics the protein environment, where the cofactors are anchored firmly in place.) These calculations were carried out using the MNDO 94 program in UNICHEM, with the AM1 Hamiltonian.

Relaxation of the active site plus the whole protein. To take into account the effects of relaxation of the protein, the cofactors and the substrate in the protein environment, quantum mechanical calculations of the active site were combined with molecular dynamics calculations of the complete system (protein plus cofactors) using procedures implemented in CHARMM[20,25]. For these combined calculations the results from the quantum chemical DFT B3LYP/6-31G(d) calculations (geometry and Mulliken charges of the cofactors and of residues around the active site of the enzyme) were input as part of the CHARMM parameter and topology files [34].

The structures including these modifications were minimized with CHARMM and used as the initial models in whole enzyme molecular dynamics calculations of the different species involved in the catalytic mechanism: QR1 with reduced N-methyl nicotinamide and oxidized FAD, QR1 with the transition state, QR1 with oxidized N-methyl nicotinamide and reduced FADH, and the final structure after completion of the charge relay. Molecular dynamics calculations were run starting with the minimized structures, using an integration step of 1 fs. The molecule was heated from 0 to 300 K, increasing the temperature by 10 K every 100 steps (total 3,100 steps; 3.1 ps), equilibrated for 1,000 steps (1 ps), and run for 30 or more picoseconds.

The time series generated by the dynamics simulation was analyzed for the following critical atom-to-atom distances: (a) the distance between C4N (C4 of nicotinamide) and N5F (N5 of flavin), i.e., the hydride transfer path; (b) the distance between FO2 (O2 of flavin) and the O of Tyr-155, important for the first step of the charge relay performed by the enzyme; and (c) the distance between the Nε of His-161 and the hydroxyl O in Tyr-155, involved in the last step of the charge relay mechanism.

RESULTS and DISCUSSION

Transition state identification and relaxation of the cofactors. In the initial search, only movements of the hydride were allowed, while keeping all other atoms fixed. These calculations included cofactors and selected amino acid residues important in the QR1 mechanism.

The sensitivity of the transition state to movement of the cofactors was analyzed by permitting relaxation of the coordinates of the carboxyamide portion of the nicotinamide. It was found that once the hydride leaves the nicotinamide, the companion hydrogen on C4N begins to enter the plane of the nicotinamide ring and collides with one of the hydrogens of the carboxyamide. The result is that in the transition state the carboxyamide plane is rotated with respect to the plane of the ring such that the NH2 moves towards the flavin. This conformation of the carboxyamide in the transition state was confirmed by performing optimizations with different dihedral angle values for the rotation of the carboxyamide plane, moving the NH2 both towards and away from the flavin.

Calculations including the charge relay mechanism. The ensuing step in the proposed mechanism of QR1 involves a charge relay. The proposed charge relay consists of the migration of a proton from Tyr-155 to O2F of the flavin, followed by a second proton transfer from His-161 to Tyr-155. It is not known whether the proton transfers take place as part of the formation of the transition state or after the hydride transfer has already occurred. To test both possibilities the energy of the structures with the transition state and either one or both protons transferred was calculated as described in Methods. Results of these calculations show that transference of one or both protons does not stabilize the transition state suggesting that the transition state is formed before either of the two protons is transferred.

Relaxation of the intermediate and final structures. To analyze the effect of cofactor relaxation the energies were recalculated allowing the cofactors to relax by optimizing their geometries with the semi-empirical MNDO 94 program in UNICHEM, with the AM1 Hamiltonian. The sugar and adenine portions of both cofactors and of the side chains included in the QM calculations were kept fixed at their x-ray positions.

Although grid calculations performed allowing only movement of the hydride show a clear saddle point (transition state), the potential energy surface is asymmetrical, indicating the development of unbalanced forces upon movement of the hydride (Fig. 1).

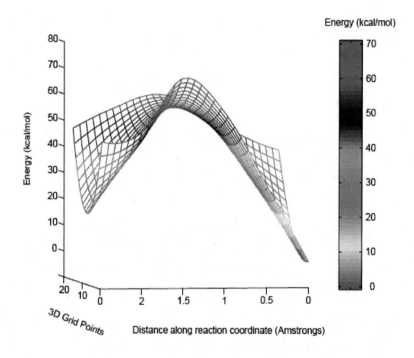

FIGURE 1. *Energy surface obtained with single point energy calculations.* The points of the 3D grid are along and around the lines joining the position of the hydride at the transition state with its positions when bound to the donor (C4 of nicotinamide) and the acceptor (N5 of the flavin).

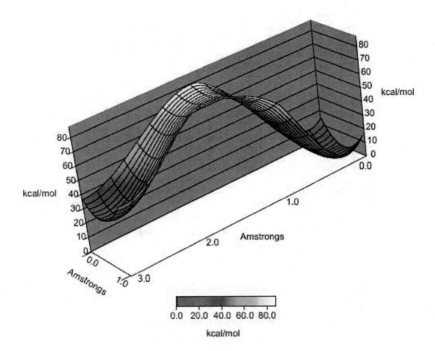

FIGURE 2. *Energy surface obtained allowing relaxation of the cofactors.* The points are defined as those in Fig. 1.

Moreover, the estimated activation energy is 80 kcal/mol, about one order of magnitude too large to account for the experimental rate of reduction of the enzyme. To reduce this strain, an optimization was performed at each grid point, allowing the cofactors to relax but fixing their sugar-adenine portions, considered to be anchored in place by the protein. The resulting energy surface (Fig. 2) clearly shows that the unbalanced forces are a consequence of restricting the movement of the cofactors: once they are allowed to relax, the resulting energy surface is symmetrical.[35-38]

An additional optimization allowing the active site side chains to move without constraints, lowered the transition state energy by about 30 kcal/mol. However, this approximation is not realistic because the side chains move to positions where the rest of the protein might not permit them to move. Allowing side chain movement by combined QM/MM methodology was then used to include the relaxation by molecular mechanics of the whole protein around the quantum mechanically-treated active site.

Enzymatic rate enhancement by relaxation of the whole protein. For these calculations the reference state was the enzyme before the reaction, with oxidized FAD and the reduced NADH lying above the FAD. Exploratory molecular dynamics runs showed that after heating to 300 K (3.1 ps), early equilibration (1.0 ps) and molecular dynamics (70.0

ps), the difference in potential energy between the initial and the transition states stabilizes once the simulation was run for at least 65 picoseconds (Fig. 3).

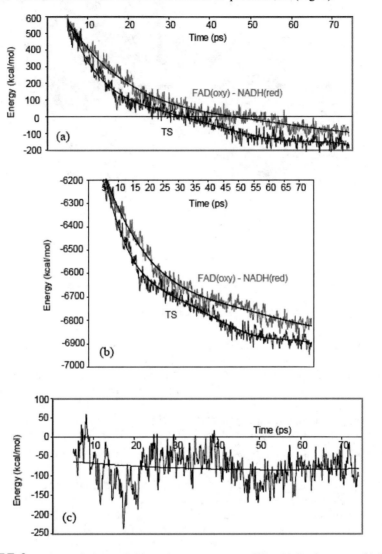

FIGURE 3. *Energies before and during the transition state (TS).* (a) Total energy. (b) Potential energy. (c)Potential energy difference between the initial sate and the transition state.

The steady state value of this difference in potential energy indicates that relaxation of the protein during the reaction lowers the transition state barrier by an average of 70 kcal/mol. Since the activation energy estimated without protein relaxation is ~80 kcal/mol, allowing for relaxation of the protein lowers the activation energy to ~10 kcal/mol (Fig. 4), a range compatible with the experimental data.

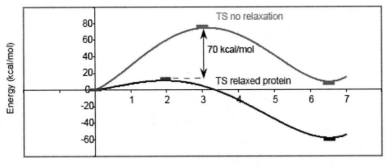

FIGURE 4. *Effect of the relaxation of the protein on the energy barrier of the reaction.* The energies are shown using the energy before the reaction as the reference state (zero value). The reduction in the activation energy (~70 Kcal/mol) is the result of allowing the protein to relax by MD simulations.

It must be noted that the molecular dynamics energies fluctuate, and thus the actual lowering of the barrier at any given instant may be more or less than the average of 70 kcal/mol. This effect is similar to the rate-promoting vibrations discussed in the analysis of the reaction of horse liver alcohol dehydrogenase[39], and to the network of coupled motions[40] and the preorganization and protein dynamics effect[2] found in hydride transfer in dihydrofolate reductase. It also similar to the contribution to catalysis of conformational fluctuations found in a study of in HIV-1 protease[32], and by acyl carrier protein reductase from *Mycobacterium tuberculosis*[33].

Molecular dynamics studies allowing relaxation of the whole protein highlight the contribution of protein flexibility to enzymatic catalysis[2,39,41]. For example, in some cases, the crystallographic distance between two atoms proposed to interact during catalysis is too large[40,41], but this difficulty may be artifactual: protein motions may bring the two atoms closer together a significant fraction of the time[41], and allow the reaction to occur. The procedure used in this study allows exploration of this type of dynamic contribution of the protein to the enzymatic mechanism of QR. The dynamic relaxation that lowers the energy of the transition state also lowers the mean atom-to-atom distances between reacting groups, favoring the overall reaction. The distance between C4N and N5F (the hydride transfer path) before the reaction occurs (red curve in Fig. 5a) is approximately 3.5 Å, goes up to more than 4.0 Å, and then returns to around 3.5 Å. However, at the transition state (blue curve in Fig. 5a) the distance between C4N and N5F becomes shorter. It is less than 3.1 Å 40% of the time of the simulation, and shorter than 2.9 Å for 15% of the time (Fig. 6a). This will no doubt facilitate the hydride transfer, because the electronic charge seems to only partially migrate at the transition state. At the transition state the hydride only has a negative charge of -0.35 rather than the expected -1.00, with the rest of the electronic charge still with the NAD. The close proximity of C4N and N5F will facilitate migration of the charge, probably through a HOMO (High Occupied Molecular Orbital) shared between NAD, the hydride and FAD.

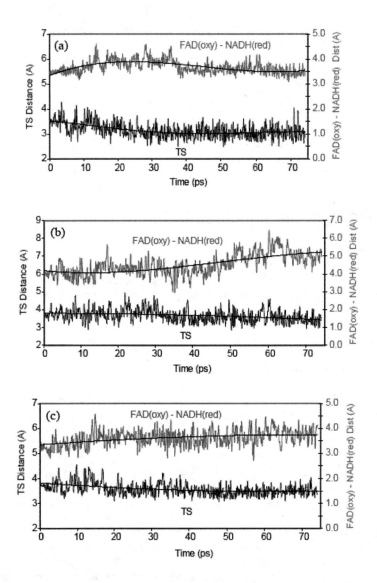

FIGURE 5. *Distances between critical atoms of the cofactors and the protein side-chains during the MD simulations.* (a) Distance between nicotinamide C4 and flavin N5. (b) Distance between His 161 Nε and Tyr 155 OH. (c) Distance between Tyr 155 OH and flavin O2. Distances in diagrams (b) and (c) correspond to the proton donors and acceptors of the charge relay. Curves corresponding to the simulation before the reaction are in red, and those in the transition state are in blue.

The short distance between C4N and N5F may even allow tunneling[2,36-38]. The distance between FO2 and the O in Tyr-155 before the reaction (red curve, Fig. 5c) begins at ~3.5

Å and increases to ~3.8 Å. At the transition state (blue curve, Fig. 5c) it begins at ~3.8 Å and eventually reaches close to hydrogen-bonding distance (~3.5 Å). This close apposition will facilitate the proton transfer from Tyr-155 to FO2, and thus facilitate the charge relay.

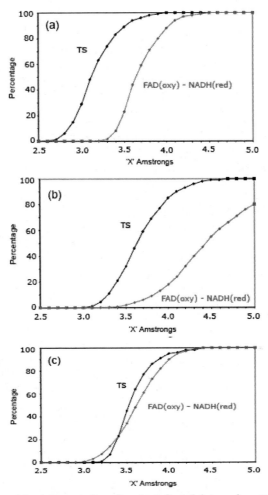

FIGURE 6. *Percentage of the time spent closer than the indicated distance by critical atoms of the cofactors and the protein side-chains during the MD simulations.* (a) Distance between nicotinamide C4 and flavin N5. (b) Distance between His 161 Nε and Tyr 155 OH. (c) Distance between Tyr 155 OH and flavin O2. Diagrams (a) and (b) correspond to the distances between the proton donors and acceptors of the charge relay. Curves corresponding to the simulation before the reaction are in red, and those in the transition state are in black.

A similar relaxation was observed in calculations performed for a similar enzymatic reaction, the deacylation in Class A Beta-Lactamases[41]. When the structure was allowed to relax, calculations showed distance shortening and energy barrier lowering for

9

different steps of the proposed reaction mechanism that should greatly facilitate the reaction. In particular, the crystallographic coordinates show a large distance between the catalytic glutamate and serine residues, excluding direct proton transfer. However, upon relaxation by QM/MM, the proposed mechanism becomes energetically feasible and the distances for proton transfer are within reasonable limits.

Before the reaction, the distance between the Nε of His-161 and the O of Tyr-155 (red curve in Fig. 5b) is around 4.0 Å, and increases steadily to close to 5.0 Å. However, at the transition state (blue curve in Fig. 5b), the simulation shows that these two atoms come within hydrogen bonding distance, thus facilitating the second step in the charge relay. The set of motions found here, which results in distance shortening and energy barrier lowering in quinone reductase, is similar to the network of coupled motions described in the hydride transfer reaction of dihydrofolate reductase[40]. Taken together all these results show that relaxation of the entire protein during the catalytic reaction can bring interacting atoms closer together and significantly lower the transition state energy, greatly facilitating the enzymatic reaction, both in energetic and in dynamic terms[40,42-44].

Extensive non-statistical dynamics have been shown to be important in gas-phase SN2 reactions[45]. The molecular dynamics simulations performed here with QM-derived atomic charges extend this type of analysis to enzyme systems and provide non-statistical details that are not accessible from studies based on (static) potential energy surfaces analyzed with transition state theory. In particular, motions of protein atoms that participate in the reaction mechanism provide an important contribution to catalysis because they result in transient close proximity of reacting groups.

An experimental evaluation of the contribution of enzyme dynamics to catalysis has been carried out for conformational fluctuations of dihydrofolate reductase in the micro- to mili-second range using nuclear magnetic resonance relaxation methods[46]. It was found that the time scales of protein dynamics coincide with those of substrate turnover, but the details of the motions during catalysis cannot be observed by this method[46]. Molecular dynamics methods such as those used here, based on quantum chemical and molecular dynamics calculation before and after the reaction, provide a simulation tool to study in atomic detail the contribution of atomic movements to catalysis[40].

SUMMARY AND CONCLUSIONS

Contributions to catalysis. The factors that increase the enzymatic reaction rate and result in rate-enhancement by QR1 can be dissected by referring to general mechanisms known to be important for rate enhancement in enzyme catalysis: *Approximation, Charge Relay, Electrostatic Catalysis, and Strain or Distortion*[47,48]. This study shows that in QR these contributions are amplified by dynamic effects, giving a high overall rate enhancement[41,49].

Approximation (effective concentration) plays an important role in catalysis by QR. The enzyme provides an optimal initial alignment of the substrate and flavin cofactor, such that the hydride needs to cross a distance of only 2.5 Å (d_{C4-N5}=3.5 Å). Dynamic effects (Fig. 5a, blue trajectory) reduce this distance by an average to 0.5 Å. (Fig. 6a shows that the NC4-FN5 distance is below 3 Å approximately 25% of the time.)

The distance between the Nε of His-161 and the OH oxygen of Tyr-155 is also diminished dynamically (Fig. 5b, blue trajectory; and Fig. 6b) by 0.3-0.8 Å. In the transition state this distance is shorter than 3.5 Å 25% of the time. A similar shortening is observed for the distance between the oxygen in Tyr-155 and the FO2 in the isoalloxazine ring (Fig. 5c, blue trajectory; and Fig. 6c). With atoms at these short distances a significant fraction of the time, the transfer of the hydride and the proton should be highly facilitated. If the rate enhancement by dynamic proximity is greater than the inverse of the fraction of the time spent at that distance, the net effect will be an increase of the rate.

It is obvious that *Electrostatic Catalysis* plays an important role in rate enhancement by QR. That this contribution is enhanced by dynamic effects can be inferred from Fig. 3. Dynamic electrostatic rearrangements in and around the active site upon going from the initial state to the transition state significantly lower the energy of the transition state.

A final effect important in rate enhancement by QR is *Strain or Distortion*. The optimizations and energy analysis indicate that the protein holds the reduced flavin in its strained planar conformation, destabilizing the ground state of the back hydride transfer reaction (from the isoalloxazine ring of the flavin to the quinone/substrate) by 2 to 4 kcal/mol. This makes the flavin a better reductant for the wide variety of quinones that it is known to reduce.

The present study also shows that, as part of these contributions, *dynamic effects* play an important role in catalysis by QR. During the hydride transfer from NADH to the FAD the hydride donor (C4 of nicotinamide) and acceptor (N5 of FAD) spend part of their time at distances that favor the formation of the transition state of the transfer. Hydride transfer takes place when, under normal thermal motion, the protein adopts one of the conformations that favors the transfer. Since these conformations are present a significant fraction of the time, transfer when the protein adopts them must represent one of the important paths in the mechanism of the reaction.

Using the procedures described here it will be straightforward to make predictions about relative rate enhancements for different quinone pro-drug substrates in the design of cancer chemotherapeutic drugs [4-6]. For example, calculations may be used to redesign drugs known to be activated by quinone reductase that are reduced more effectively by the rat enzyme than by human QR[50].

ACKNOWLEDGMENTS

Supported by Grant GM 51362 of the National Institute of General Medical Sciences. We thank the Supercomputing Center of the National Cancer Institute, Frederick, Md.

REFERENCES

1. Cavelier G, Amzel LM. Mechanism of NAD(P)H : quinone reductase: Ab initio studies of reduced flavin. Proteins-Structure Function and Genetics 2001;43(4):420-432.
2. Rajagopalan PT, Benkovic SJ. Preorganization and protein dynamics in enzyme catalysis. Chem Rec 2002;2(1):24-36.
3. Frisch MJ, Trucks GW, Schlegel HB, Scuseria GE, Robb MA, Cheeseman JR, Zakrzewski VG, Montgomery JA, Stratmann RE, Burant JC, Dapprich S, Millam JM, Daniels AD, Kudin KN, Strain MC, Farkas O, Tomasi J, Barone V, Cossi M, Cammi R, Mennucci B, Pomelli C, Adamo C, Clifford S, Ochterski J, Petersson GA, Ayala PY, Cui Q, Morokuma K, D. K. Malick, Rabuck AD, Raghavachari K, Foresman JB, Cioslowski J, J. V. Ortiz, Stefanov BB, Liu G, Liashenko A, Piskorz P, Komaromi I, R. Gomperts, Martin RL, Fox DJ, Keith T, Al-Laham MA, Peng CY, Nanayakkara A, Gonzalez C, Challacombe M, Gill PMW, Johnson BG, Chen W, Wong MW, Andres JL, Head-Gordon M, an ESR, Pople JA. Gaussian 98. Pittsburgh PA: Gaussian, Inc.; 1998. 280 p. p.
4. Amzel LM. Structure-based drug design. Current Opinion in Biotechnology 1998;9(4):366-369.
5. Faig M, Bianchet MA, Talalay P, Chen S, Winski S, Ross D, Amzel LM. Structures of recombinant human and mouse NAD(P)H : quinone oxidoreductases: Species comparison and structural changes with substrate binding and release. Proceedings of the National Academy of Sciences of the United States of America 2000;97(7):3177-3182.
6. Faig M, Bianchet MA, Winski S, Hargreaves R, Moody CJ, Hudnott AR, Ross D, Amzel LM. Structure-based development of anticancer drugs: Complexes of NAD(P)H : quinone oxidoreductase 1 with chemotherapeutic quinones. Structure 2001;9(8):659-667.
7. Li R, Bianchet MA, Talalay P, Amzel LM. The three-dimensional structure of NAD(P)H:quinone reductase, a flavoprotein involved in cancer chemoprotection and chemotherapy: Mechanism of the two-electron reduction. Proc Natl Acad Sci USA 1995;92:8846-8850.
8. Abashkin Y, Russo N, Toscano M. Transition-State Localization By a Density-Functional Method - Applications to Isomerization and Symmetry-Forbidden Reactions. Theoretica Chimica Acta 1995;91(3-4):179-186.
9. Alhambra C, Corchado JC, Sanchez ML, Gao JL, Truhlar DG. Quantum dynamics of hydride transfer in enzyme catalysis. J Am Chem Soc 2000;122(34):8197-8203.
10. Andres J, Moliner V, Safont VS, Domingo LR, Picher MT. On Transition Structures for Hydride Transfer Step in Enzyme Catalysis. A Comparative Study on Models of Glutathione Reductase Derived from Semiempirical, HF, and DFT Methods. J Org Chem 1996;61(22):7777-7783.
11. Andres J, Moliner V, Safont VS, Aullo JM, Diaz W, Tapia O. Transition structures for hydride transfer reactions in vacuo and their role in enzyme catalysis. Theochem-Journal of Molecular Structure 1996;371:299-312.
12. Hurley MM, HammesSchiffer S. Development of a potential surface for simulation of proton and hydride transfer reactions in solution: Application to

NADH hydride transfer. Journal of Physical Chemistry a 1997;101(21):3977-3989.

13. Diaz W, Aullo JM, Paulino M, Tapia O. Transition structure and reactive complexes for hydride transfer in an isoalloxazine-nicotinamide complex. On the catalytic mechanism of glutathione reductase. An ab initio MO SCF study. Chemical Physics 1996;204(2-3):195-203.

14. Ferenczy GG, Naray-Szabo G, Varnai P. Quantum mechanical study of the hydride shift step in the xylose isomerase catalytic reaction with the fragment self- consistent field method. International Journal of Quantum Chemistry 1999;75(3):215-222.

15. Mestres J, Duran M, Bertran J. Characterization of the Transition State for the Hydride Transfer in a Model of the Flavoprotein Reductase Class of Enzymes. Bioorg Chem 1996;24:69-80.

16. Nishimoto K, Higashimura K, Asada T. Ab initio molecular orbital study of the flavin-catalyzed dehydrogenation reaction of glycine - protein transport channel driving hydride-transfer mechanism. Theoretical Chemistry Accounts 1999;102(1-6):355-365.

17. Park BK, Doh ST, Son GS, Kim JM, Lee GY. Mo Study of Hydride Transfer Between Nadh and Flavin Nucleotides. Bulletin of the Korean Chemical Society 1994;15(4):291-293.

18. Webb SP, Agarwal PK, Hammes-Schiffer S. Combining electronic structure methods with the calculation of hydrogen vibrational wavefunctions: Application to hydride transfer in liver alcohol dehydrogenase. Journal of Physical Chemistry B 2000;104(37):8884-8894.

19. Tapia O, Andrés J, Safont VS. Transition structures in vacuo and the theory of enzyme catalysis. Rubisco's catalytic mechanism: a paradigmatic case? Journal of Molecular Structure: THEOCHEM 1995;342(1-3):131-140.

20. Brooks BR, Bruccoleri RE, Olafson BD, States DJ, Swaminathan S, Karplus M. Charmm - a Program For Macromolecular Energy, Minimization, and Dynamics Calculations. Journal of Computational Chemistry 1983;4(2):187-217.

21. Cui Q, Karplus M. Molecular properties from combined QM/MM methods. I. Analytical second derivative and vibrational calculations. J Chem Phys 2000;112(3):1133-1149.

22. Cummins PL, Gready JE. Molecular dynamics and free energy perturbation study of hydride-ion transfer step in dihydrofolate reductase using combined quantum and molecular mechanical model. Journal of Computational Chemistry 1998;19(8):977-988.

23. Liu H, Müller-Plathe F, Gunsteren WFv. A Combined Quantum/Classical Molecular Dynamics Study of the Catalytic Mechanism of HIV Protease. Journal of Molecular Biology 1996;261(3):454-469.

24. Sheppard DW, Burton NA, Hillier IH. Ab initio hybrid quantum mechanical/molecular mechanical studies of the mechanisms of the enzymes protein kinase and thymidine phosphorylase. Journal of Molecular Structure-Theochem 2000;506:35-44.

25. Cui Q, Karplus M. Triosephosphate isomerase: a theoretical comparison of alternative pathways. J Am Chem Soc 2001;123(10):2284-2290.

26. Ramos MJ, Melo A, Henriques ES, Gomes J, Reuter N, Maigret B, Floriano WB, Nascimento MAC. Modeling enzyme-inhibitor interactions in serine proteases. International Journal of Quantum Chemistry 1999;74(3):299-314.

27. Varnai P, Warshel A. Computer simulation studies of the catalytic mechanism of human aldose reductase. J Am Chem Soc 2000;122(16):3849-3860.

28. Bentzien J, Muller RP, Florian J, Warshel A. Hybrid ab initio quantum mechanics molecular mechanics calculations of free energy surfaces for enzymatic reactions: The nucleophilic attack in subtilisin. Journal of Physical Chemistry B 1998;102(12):2293-2301.

29. Kong YS, Warshel A. Linear Free-Energy Relationships With Quantum-Mechanical Corrections - Classical and Quantum-Mechanical Rate Constants For Hydride Transfer Between Nad(+) Analogs in Solutions. J Am Chem Soc 1995;117(23):6234-6242.

30. Yadav A, Jackson RM, Holbrook JJ, Warshel A. Role of Solvent Reorganization Energies in the Catalytic Activity of Enzymes. J Am Chem Soc 1991;113(13):4800-4805.

31. Cavalli A, Carloni P. Enzymatic GTP hydrolysis: insights from an ab initio molecular dynamics study. J Am Chem Soc 2002;124(14):3763-3768.

32. Piana S, Carloni P, Parrinello M. Role of conformational fluctuations in the enzymatic reaction of HIV-1 protease. J Mol Biol 2002;319(2):567-583.

33. Pantano S, Alber F, Lamba D, Carloni P. NADH interactions with WT- and S94A-acyl carrier protein reductase from Mycobacterium tuberculosis: an ab initio study. Proteins 2002;47(1):62-68.

34. Pavelites JJ, Gao JL, Bash PA, Mackerell AD. A molecular mechanics force field for NAD(+), NADH, and the pyrophosphate groups of nucleotides. Journal of Computational Chemistry 1997;18(2):221-239.

35. Foresman JB, Frisch A. Exploring Chemistry with Electronic Structure Methods. Pittsburgh, PA: Gaussian, Inc.; 1996. 301 p. p.

36. Ringe D, Petsko GA. Tunnel vision. Nature 1999;399(3 June 1999):417-418.

37. Kohen A, Cannio R, Bartolucci S, Klinman JP. Enzyme dynamics and hydrogen tunneling in a thermophilic alcohol dehydrogenase. Nature 1999;399(3 June 1999):496-499.

38. Kohen A, Klinman JP. Enzyme Catalysis: Beyond Classical Paradigms. Accounts Chem Res 1998;31:397-404.

39. Caratzoulas S, Mincer JS, Schwartz SD. Identification of a protein-promoting vibration in the reaction catalyzed by horse liver alcohol dehydrogenase. J Am Chem Soc 2002;124(13):3270-3276.

40. Agarwal PK, Billeter SR, Rajagopalan PT, Benkovic SJ, Hammes-Schiffer S. Network of coupled promoting motions in enzyme catalysis. Proc Natl Acad Sci U S A 2002;99(5):2794-2799.

41. Castillo R, Silla E, Tunon I. Role of protein flexibility in enzymatic catalysis: quantum mechanical- molecular mechanical study of the deacylation reaction in class A beta- lactamases. J Am Chem Soc 2002;124(8):1809-1816.

42. Bahar I, Erman B, Jernigan RL, Atilgan AR, Covell DG. Collective motions in HIV-1 reverse transcriptase: examination of flexibility and enzyme function. J Mol Biol 1999;285(3):1023-1037.

43. Keskin O, Jernigan RL, Bahar I. Proteins with similar architecture exhibit similar large-scale dynamic behavior. Biophys J 2000;78(4):2093-2106.

44. Mesecar AD, Stoddard BL, Koshland DE, Jr. Orbital steering in the catalytic power of enzymes: small structural changes with large catalytic consequences. Science 1997;277(5323):202-206.

45. Sun L, Song K, Hase WL. A SN2 reaction that avoids its deep potential energy minimum. Science 2002;296(5569):875-878.

46. Eisenmesser EZ, Bosco DA, Akke M, Kern D. Enzyme dynamics during catalysis. Science 2002;295(5559):1520-1523.

47. Garcia-Viloca M, Gao J, Karplus M, Truhlar DG. How enzymes work: analysis by modern rate theory and computer simulations. Science 2004;303(5655):186-195.

48. Strajbl M, Shurki A, Kato M, Warshel A. Apparent NAC effect in chorismate mutase reflects electrostatic transition state stabilization. J Am Chem Soc 2003;125(34):10228-10237.

49. Sulpizi M, Schelling P, Folkers G, Carloni P, Scapozza L. The rational of catalytic activity of herpes simplex virus thymidine kinase. a combined biochemical and quantum chemical study. J Biol Chem 2001;276(24):21692-21697.

50. Chen S, Knox R, Wu K, Deng PS, Zhou D, Bianchet MA, Amzel LM. Molecular basis of the catalytic differences among DT-diaphorase of human, rat, and mouse. J Biol Chem 1997;272(3):1437-1439.

Ligand Binding, Protein Fluctuations, And Allosteric Free Energy

Michael E. Wall

Computer and Computational Sciences Division & Bioscience Division, Los Alamos National Laboratory, Los Alamos, NM 87545 USA. E-mail: mewall@lanl.gov

Abstract. Although the importance of protein dynamics in protein function is generally recognized, the role of protein fluctuations in allosteric effects scarcely has been considered. To address this gap, the Kullback-Leibler divergence (D_x) between protein conformational distributions before and after ligand binding was proposed as a means of quantifying allosteric effects in proteins. Here, previous applications of D_x to methods for analysis and simulation of proteins are first reviewed, and their implications for understanding aspects of protein function and protein evolution are discussed. Next, equations for D_x suggest that $k_B T D_x$ should be interpreted as an allosteric free energy – the free energy associated with changing the ligand-free protein conformational distribution to the ligand-bound conformational distribution. This interpretation leads to a thermodynamic model of allosteric transitions that unifies existing perspectives on the relation between ligand binding and changes in protein conformational distributions. The definition of D_x is used to explore some interesting mathematical relations among commonly recognized thermodynamic and biophysical quantities, such as the total free energy change upon ligand binding, and ligand-binding affinities for individual protein conformations. These results represent the beginnings of a theoretical framework for considering the full protein conformational distribution in modeling allosteric transitions. Early applications of the framework have produced results with implications both for methods for coarsed-grained modeling of proteins, and for understanding the relation between ligand binding and protein dynamics.

Keywords: Kullback-Leibler divergence, protein conformational distribution, protein dynamics, protein vibrations, allostery, coarse-grained model, molecular engine
PACS: 87.15.-v
Technical Release Number: Los Alamos National Laboratory LA-UR-06-2066

INTRODUCTION

One important mechanism of protein regulation is allosteric regulation, in which molecular interactions influence protein activity through changes in protein structure. In traditional models of allosteric regulation, proteins adopt a limited number of conformations, each of which may have a different activity [1,2]. However, the importance of considering continuous conformational distributions in understanding allosteric effects was recognized by Weber [3]. Neutron scattering experiments later provided evidence for changes in protein dynamics upon ligand binding [4], and it was subsequently realized that ligand binding at an allosteric site can influence binding at a remote site without inducing a mean conformational change, solely through alteration

CP851, *From Physics to Biology; BIFI 2006 II International Congress,*
edited by J. Clemente-Gallardo, Y. Moreno, J. F. Sáenz Lorenzo, and A. Velázquez-Campoy
© 2006 American Institute of Physics 0-7354-0350-3/06/$23.00

of atomic fluctuations [5]. Indeed, the conformational distribution is known to be a key determinant of protein activity [6], and is a key element in rate theories [7].

Motivated by these considerations, a theoretical framework was recently developed to quantify changes in protein conformational distributions upon ligand binding in terms of the Kullback-Leibler divergence [8], D_x [9,10]. A closed-form estimate of D_x was derived in the harmonic approximation [10], and was used to demonstrate that values of D_x are elevated in small-molecule binding sites of proteins [11]. To analyze aspects of allosteric mechanisms, D_x was calculated for local regions of bovine trypsinogen upon binding ligands in the active site or allosteric site of the protein. In addition, D_x has been used to develop rigorous methods for evaluating and optimizing coarse-grained models of proteins [10]. These previous results are reviewed below.

While previous work has focused on the application of D_x to methods for analysis and simulation of proteins, the biophysical meaning of D_x has not yet been described in detail. Therefore, after the review of previous results is a discussion of aspects of the thermodynamic and biophysical significance of D_x, with implications for understanding allosteric effects in proteins.

DEFINITION OF D_X

The Kullback-Leibler divergence D_x upon binding a ligand is defined as

$$D_x = \int d^{3N}\mathbf{x}\ P'(\mathbf{x})\ln\frac{P'(\mathbf{x})}{P(\mathbf{x})} \qquad (1)$$

where $P'(\mathbf{x})$ is the protein conformational distribution in the presence of the ligand, and $P(\mathbf{x})$ is the same distribution in the absence of the ligand [9]. It is easily shown that D_x is always non-negative. Note that D_x is not a distance measure because its value is not conserved when $P'(\mathbf{x})$ and $P(\mathbf{x})$ are interchanged in Eq. (1).

Let a reaction rate $k(\mathbf{x})$ be a function of the configuration $\mathbf{x}$ of N atoms. The relation between the rate distribution $P(k)$ and the conformational distribution $P(\mathbf{x})$ is then [9]

$$P(k) = \int d^{3N}\mathbf{x}P(\mathbf{x})\delta[k(\mathbf{x}) - k], \qquad (2)$$

where the integral is over all conformations $\mathbf{x}$. Equation (2) concisely expresses the necessity of considering the full conformational distribution in understanding the functional consequences of allosteric transitions.

D_X IN THE HARMONIC APPROXIMATION

To calculate D_x using Eq. (1), it is necessary to calculate the marginal probability distribution of the protein configurations $P(\mathbf{x})$ from a full conformational distribution $P(\mathbf{x},\mathbf{y})$ of a protein-ligand complex,

$$P(\mathbf{x}) = \int d^{3N_y}\mathbf{y}P(\mathbf{x},\mathbf{y}). \qquad (3)$$

In Eq. (3), $\mathbf{y}$ is a vector of the N_y ligand coordinates. Solutions for Equations (3) and (1) have been derived in a model of harmonic vibrations of the protein-ligand system

17

[9,10]. Let $\mathbf{z} = (\mathbf{x},\mathbf{y})$ be the coordinates of the combined protein-ligand system, measured relative to an equilibrium configuration $\mathbf{z}_0 = (\mathbf{x}_0,\mathbf{y}_0)$. Consider a harmonic approximation to the potential energy function $U(\mathbf{z}_0 + \mathbf{z})$,

$$U(\mathbf{z}_0 + \mathbf{z}) \approx U(\mathbf{z}_0) + \frac{1}{2}\mathbf{z}^\mathrm{T}\mathbf{Hz}, \tag{4}$$

where $\mathbf{H}$ is the Hessian of U evaluated at $\mathbf{z}_0$: $H_{ij}\big|_{\mathbf{z}_0} = \partial^2 U/\partial z_i \partial z_j \big|_{\mathbf{z}_0}$. Assuming a Boltzmann distribution for $P(\mathbf{z})$ and ignoring solvent and pressure effects,

$$\begin{aligned} P(\mathbf{z}) &= Z^{-1} e^{-\mathbf{z}^\mathrm{T}\mathbf{Hz}/2k_B T} \\ &= (2\pi k_B T)^{-3N_z/2} e^{-\left|\mathbf{\Omega V}^\mathrm{T}\mathbf{z}\right|^2/2k_B T} \prod_{\substack{i=1...3N_z}}^{\omega_i \neq 0} \omega_i, \end{aligned} \tag{5}$$

where Z is the partition function, k_B is Boltzmann's constant, T is the temperature, N_z is the number of atoms in the complex, the elements of the matrix $\Omega^2 = diag(\omega_1^2,...,\omega_{3N_z}^2)$ are the eigenvalues of $\mathbf{H}$, and the columns of the matrix $\mathbf{V}$ are the eigenvectors of $\mathbf{H}$. Here and elsewhere, products and summations are carried out over nonzero modes. Define the submatrices $\mathbf{H_x}$, $\mathbf{H_y}$, and $\mathbf{G}$ as

$$\mathbf{Hz} = \begin{pmatrix} \mathbf{H_x} & \mathbf{G} \\ \mathbf{G}^T & \mathbf{H_y} \end{pmatrix}\begin{pmatrix} \mathbf{x} \\ \mathbf{y} \end{pmatrix} = \begin{pmatrix} \mathbf{H_x x + G y} \\ \mathbf{G}^T\mathbf{x + H_y y} \end{pmatrix}. \tag{6}$$

$\mathbf{H_x}$ couples coordinates from $\mathbf{x}$, $\mathbf{H_y}$ couples coordinates from $\mathbf{y}$, and $\mathbf{G}$ couples coordinates between $\mathbf{x}$ and $\mathbf{y}$. By re-expressing $\left|\mathbf{\Omega V}^\mathrm{T}\mathbf{z}\right|^2$ in Eq. (5) using Eq. (6), the integral in Eq. (3) may be performed, yielding

$$P(\mathbf{x}) = (2\pi k_B T)^{-3N/2} e^{-\left|\overline{\mathbf{\Omega}}\overline{\mathbf{V}}^\mathrm{T}\mathbf{z}\right|^2/2k_B T} \prod_{\substack{i=1...3N}}^{\overline{\omega}_i \neq 0} \overline{\omega}_i. \tag{7}$$

In Eq. (7), the elements of the matrix $\overline{\Omega}^2 = diag(\overline{\omega}_1^2,...,\overline{\omega}_{3N}^2)$ and the columns $\overline{\mathbf{v}}_i$ of the matrix $\overline{\mathbf{V}}$ are the eigenvalues and eigenvectors of a matrix $\overline{\mathbf{H}}$, defined as

$$\overline{\mathbf{H}} = \mathbf{H_x} - \mathbf{GH_y^{-1}G^T} = \overline{\mathbf{V}}\left|\overline{\mathbf{\Omega}}\right|^2\overline{\mathbf{V}}^\mathrm{T}. \tag{8}$$

Equation (8) was independently derived by Ming & Wall [10] and by Zheng & Brooks [12]. When there is no interaction between the protein and ligand coordinates, Eq. (7) is just the conformational distribution of the protein in the absence of the ligand; let the unprimed probability distribution $P(\mathbf{x})$ in Eq. (7) correspond to this case. When there is an interaction between the protein and the ligand, let the distribution $P'(\mathbf{x})$ be given by a similar expression,

$$P'(\mathbf{x}) = (2\pi k_B T)^{-3N/2} e^{-\left|\overline{\mathbf{\Omega}}'\overline{\mathbf{V}}'^\mathrm{T}\mathbf{z}\right|^2/2k_B T} \prod_{\substack{i=1...3N}}^{\overline{\omega}_i' \neq 0} \overline{\omega}_i'. \tag{9}$$

With respect to the unprimed variables in Eq. (7), the primed variables in Eq. (9) correspond to the case in which the protein and ligand interact. Using methods described in detail in Ref. [9], substituting Eqs. (7) and (9) in Eq. (1) yields the following expression for $D_\mathbf{x}$:

$$D_x = \sum_{i=1\ldots3N}^{\overline{\omega}_i' \neq 0; \overline{\omega}_i \neq 0} \left(\log\frac{\overline{\omega}_i'}{\overline{\omega}_i} + \frac{1}{2k_BT}\overline{\omega}_i'^2\left|\Delta\mathbf{x}_0 \cdot \overline{\mathbf{v}}_i\right|^2 + \frac{1}{2}\sum_{j=1\ldots3N}^{\overline{\omega}_j \neq 0} \frac{\overline{\omega}_j^2}{\overline{\omega}_i'^2}\left|\overline{\mathbf{v}}_i' \cdot \overline{\mathbf{v}}_j\right|^2 - \frac{1}{2} \right). \quad (10)$$

The first term is proportional to the entropy change upon releasing the ligand, and the second term is proportional to the potential energy required to change the mean conformation of the protein without the ligand to that with the ligand. Equation (10) and variants thereof have been used in several practical applications [10,11], which are reviewed in the next section.

PRACTICAL APPLICATIONS OF D_X

Prediction Of Ligand-Binding Sites

Recently Ming & Wall [11] examined 305 protein structures from the GOLD docking test set [13] and investigated whether interactions at small-molecule binding sites cause a large change in the protein conformational distribution. A computational method, called dynamics perturbation analysis (DPA), was presented to identify sites at which interactions yield a large value of D_x. DPA was used to analyze proteins in the test set, and to determine whether D_x values for points in the neighborhood of ligand-binding sites were high compared to random points. A method was then developed to predict functional sites in proteins, and the method was evaluated using proteins in the test set. The performance of the method was compared to that of a cleft analysis method.

The DPA method was based on a method previously used to analyze changes in molecular vibrations of a lysozyme-NAG complex for random protein-ligand interactions [9]. In DPA, a protein is decorated with M surface points that interact with neighboring protein atoms. The protein conformational distribution $P^{(0)}(\mathbf{x})$ is calculated in the absence of any surface points, and M protein conformational distributions $P^{(m)}(\mathbf{x})$ are calculated for the protein interacting with each point m. The conformational distributions are calculated using a coarse-grained model of molecular vibrations, and the distributions $P^{(m)}(\mathbf{x})$ are calculated from models of the protein in complex with each surface point using Eq. (9). The Kullback-Leibler divergence $D_x^{(m)}$ between $P^{(0)}(\mathbf{x})$ and $P^{(m)}(\mathbf{x})$ is calculated for each point m (Eq. (10)), and is used as a measure of the change in the protein conformational distribution upon interacting with point m.

Protein vibrations were modeled using the elastic network model (ENM) [14-17]. In the ENM, alpha-carbon atoms are extracted from an atomic model of a protein, and an interaction network is generated by connecting springs between all atom pairs separated by a distance less than or equal to a cutoff distance r_c. Each spring has the same force constant γ, is aligned with the separation between the connected atoms, and has an equilibrium length equal to the equilibrium distance between the atoms. The interaction between the protein and a surface point m was modeled by connecting springs of force constant γ_s between the surface point and all protein atoms within a cutoff distance r_s of the surface point. The protein coordinates were not modified in modeling the interaction.

Consistent with results obtained using an all-atom model of lysozyme [9], values of $D_x^{(m)}$ were found to be elevated in the neighborhood of the tri-N-acetyl-D-glucosamine (tri-NAG) binding site. Interestingly, the distribution of $y = D_x^{(m)}$ values was empirically well-fit by a probability density $\rho(y)$ given by

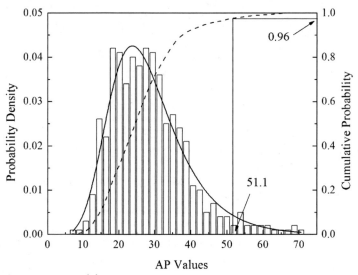

FIGURE 1. Distribution of $D_x^{(m)}$ values (labeled as AP values) for 4859 points on the surface of lysozyme (the number of points was increased in this case to evaluate the fit). The distribution is well-fit by an extreme value distribution with parameters μ = 23.07 and β = 8.45 (Pearson correlation coefficient of 0.992). The fit is used to find the 96% upper bound of $D_x^{(m)}$ for the surface points; this bound is used as the threshold to select high-$D_x^{(m)}$ points for use in predicting functional sites.

$$\rho(y) = \frac{1}{\beta} e^{\frac{y-\mu}{\beta} - e^{\frac{y-\mu}{\beta}}}. \tag{11}$$

which is an extreme value distribution of width β centered on μ (Fig. 1).

DPA was applied to 305 protein structures in the GOLD docking test set [13]. Calculations were performed in the same manner as for lysozyme. A statistical analysis method was developed to quantitatively assess the tendency for $D_x^{(m)}$ values to be elevated in the neighborhood of ligand-binding sites. In this method, surface points in the neighborhood of the ligand were selected, each point was ranked with respect to all other surface points in terms of the value of $D_x^{(m)}$, and a composite score for all points was calculated. The probability of obtaining the composite score by randomly selecting points on the surface was calculated and was used to calculate a P-value for evaluating the significance of the score for each protein. In 14 of the 305 proteins, the ligand was buried and was not close to any of the surface points. Results for the rest of the proteins are illustrated in Fig. 2. For 95% of proteins, the P-value is 10^{-3} or lower,

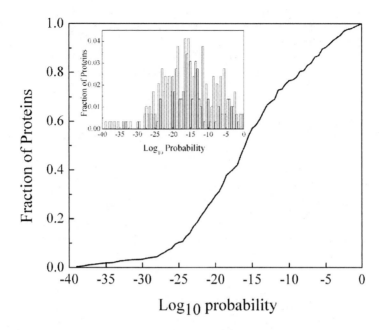

FIGURE 2. Statistical significance of elevated values of $D_x^{(m)}$ in functional sites. The distribution of P-values (calculated in bins of width 2 in log units) calculated for the composite $D_x^{(m)}$ score is shown for 291 proteins in the GOLD docking test set. For 95% of proteins, the P-value is 10^{-3} or lower, indicating that the elevation of $D_x^{(m)}$ in the neighborhood of functional sites is statistically significant.

indicating that the elevation of $D_x^{(m)}$ in the neighborhood of functional sites was statistically significant.

Following the weaker suggestion from all-atom study of lysozyme [9], these results strongly suggested that points with high values of $D_x^{(m)}$ could be used to predict the locations of functional sites, and motivated the development of an algorithm for this purpose. The algorithm works as follows. First, DPA is performed on a protein. Then, the statistics of $D_x^{(m)}$ values is modeled using an extreme value distribution. Points with significantly high values of $D_x^{(m)}$ are selected and are spatially clustered. The clusters are ranked according to the mean value of $D_x^{(m)}$ within the cluster, and points in the highest-ranked cluster are predicted to be associated with a functional site. Finally, residues in the neighborhood of the highest-ranked cluster are selected and are predicted to reside within the functional site.

Consistent with the analysis of lysozyme, the $D_x^{(m)}$ values for the test-set proteins indicated that the statistics are well-described by an extreme-value distribution. To select points with significantly high values of $y = D_x^{(m)}$, an operating point C of the cumulative distribution $c(y) = 1 - e^{-e^{\frac{y-\mu}{\beta}}}$ was selected, μ and β were fitted using the actual distribution of $D_x^{(m)}$ for the protein, and a lower threshold Y on $D_x^{(m)}$ was calculated. Points with $D_x^{(m)} > Y$ were clustered spatially, and the mean value of $D_x^{(m)}$

for each cluster was calculated and was used to rank the clusters; for each protein, the rank-1 cluster was identified as the cluster with the highest mean value.

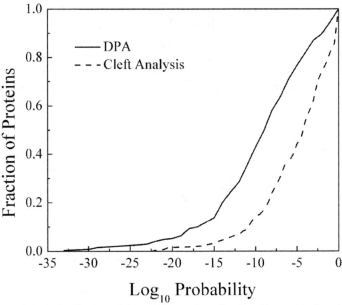

FIGURE 3. Statistical significance of the overlaps of predicted residues with ligand-binding-site residues. For each protein, a P-value (corresponding to the probability in a null model of finding at least as many ligand-binding-site resides as does the prediction algorithm) is calculated; the resulting distribution of P-values is shown here. For the DPA algorithm (solid line), a total of 250 proteins in the test set were considered; and for the cleft analysis algorithm (dashed line), a total of 278 proteins were considered. (In each case, only proteins for which the algorithm yielded at least one residue in the ligand-binding site were considered.)

Protein alpha-carbons within 6 Å of any of the points in the rank-1 cluster were selected and were used to identify the set of residues that are predicted to reside in a functional site. To evaluate the predictions, they were compared with the set of residues that are in the neighborhood of the ligand found in complex with the protein in the test set.

The method produced predictions for 287 of the 305 proteins. In 87% of cases (250 proteins), at least one predicted residue was in the ligand-binding site. The recall was at least 0.3 for 80% of cases, and was at least 0.5 for 76% of the cases. The precision was at least 0.3 for 68% of the cases, and was at least 0.5 for 44% of the cases. The statistical significance of the overlaps was assessed using a null model in which surface residues were randomly selected. Using the null model, a P-value was calculated to evaluate predictions for the 250 proteins in which at least one predicted residue was in the ligand-binding site. Results are shown in Fig. 3. For 87% of the cases, the P-value is 10^{-3} or smaller, indicating that there is a statistically significant overlap. As shown in Fig. 3, the performance of the DPA method compared favorably to that of a cleft analysis method for predicting ligand-binding residues.

Optimization Of Coarse-Grained Molecular Models

In one common coarse-graining method, an all-atom model is simplified by considering effective interactions among a subset of the atoms (e.g., just the alpha-carbons). The usual criterion for model accuracy is the ability of a model to reproduce atomic mean-squared displacements (MSDs). However, MSDs are just one aspect of protein dynamics -- a stricter criterion for the accuracy of a coarse-grained model is the similarity between the configurational distributions of the selected atoms in the coarse-grained and all-atom models. Such a criterion is also biologically relevant, in part because the conformational distribution is a key determinant of protein activity [6,7,18] (Eq. (2)).

One useful measure of the difference between conformational distributions obtained from all-atom and coarse-grained simulations is the Kullback-Leibler divergence D_x [10]. Let $\mathbf{x}_\alpha$ be the coordinates of the N_α alpha-carbons in either an all-atom model of molecular vibrations, or a coarse-grained model with parameters Γ. Define the optimal coarse-grained model as the one for which D_x between $P^{(\Gamma)}(\mathbf{x}_\alpha)$ and $P(\mathbf{x}_\alpha)$ is minimal, i.e., for which Γ is chosen such that

$$D_{\mathbf{x}_\alpha}^{(\Gamma)} = \sum_{i=1...3N_\alpha}^{\omega_i^{(\Gamma)} \neq 0; \overline{\omega}_i \neq 0} \left(\log \frac{\omega_i^{(\Gamma)}}{\overline{\omega}_i} + \frac{1}{2k_BT}\overline{\omega}_i^2 \left| \Delta \mathbf{x}_{\alpha,0} \cdot \overline{\mathbf{v}}_i \right|^2 \right.$$
$$\left. + \frac{1}{2} \sum_{j=1...3N_\alpha}^{\overline{\omega}_j \neq 0} \frac{\overline{\omega}_j^2}{\omega_i^{(\Gamma)2}} \left| \mathbf{v}_i^{(\Gamma)} \cdot \overline{\mathbf{v}}_j \right|^2 - \frac{1}{2} \right) \tag{12}$$

is minimal. In Eq. (12), $\omega_i^{(\Gamma)2}$ and $\mathbf{v}_i^{(\Gamma)}$ are the eigenvalue and eigenvector of mode i of the coarse-grained model; $\overline{\omega}_j^2$ and $\overline{\mathbf{v}}_j$ are the ith eigenvalue and eigenvector of the matrix $\overline{\mathbf{H}}$ calculated for the alpha-carbon atoms of the all-atom model (Eq. (8)), and $\Delta \mathbf{x}_{\alpha,0}$ is the difference between the equilibrium coordinates of the coarse-grained and all-atom models.

Ming & Wall [10] used Eq. (12) to find optimal solutions to the elastic network model (ENM) of protein vibrations [14,17], in which interacting alpha-carbons are connected by springs aligned with the direction of atomic separation. Following the Tirion model [17], each spring has the same force constant γ. For a given interaction network, the eigenvectors $\mathbf{v}_i^{(\Gamma)}$ are independent of γ, and each eigenvalue $\omega_i^{(\Gamma)2}$ is proportional to γ. Assuming no difference in the mean conformation between the models, the value of γ at which $D_{\mathbf{x}_\alpha}^{(\Gamma)}$ is minimal was calculated as

$$\gamma = \frac{1}{3N_\alpha} \sum_{i=1...3N_\alpha}^{a_i \neq 0} \sum_{j=1...3N_\alpha}^{\overline{\omega}_j \neq 0} \frac{\overline{\omega}_j^2}{a_i^2} \left| \mathbf{v}_i^{(\Gamma)} \cdot \overline{\mathbf{v}}_j \right|^2, \tag{13}$$

where the constants $a_i^2 = \omega_i^{(\Gamma)2}/\gamma$ are independent of γ. Interestingly, the third and fourth terms of Eq. (12) cancel for this value of γ. Therefore, the optimal coarse-grained model is one that uses an energy scale for interactions that eliminates the contribution to $D_{\mathbf{x}_\alpha}^{(\Gamma)}$ due to differences in the eigenvectors, and which, given this energy scale, maximizes the entropy of the conformational distribution $P^{(\Gamma)}(\mathbf{x}_\alpha)$.

23

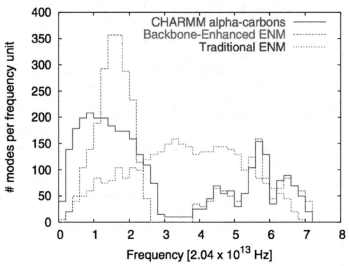

FIGURE 4. Density-of-states distribution for all-atom and elastic network models of trypsinogen. Frequency units are (Kcal / mol Å^2 m_p)$^{1/2}$ = 2.04 × 10^{13} Hz, where m_p is the proton mass. Densities were estimated by counting the number of modes in bins of width 0.2, and normalizing the integral to 663, which is the total number of non-zero modes. The ENM (dotted line) does not reproduce the bimodal distribution from the all-atom model (solid line); however, the BENM recovers the bimodal distribution (dashed line).

When the above method was applied to optimize an ENM of trypsinogen, the value of $D_{\mathbf{x}_\alpha}^{(\mathrm{r})} = 313$ was quite high [10]. This finding led to the discovery of a discrepancy between the density of states distributions from the ENM and the all-atom model: whereas the all-atom model yielded a bimodal distribution, the ENM yielded a unimodal distribution (Fig. 4). To correct this discrepancy, a backbone-enhanced elastic network model (BENM) was proposed, in which the spring constant was increased for alpha carbons that are backbone neighbors. The optimal BENM recovered the correct bimodal distribution of the density of states, and yielded a much lower value of $D_{\mathbf{x}_\alpha}^{(\mathrm{r})} = 102$. Because the agreement at high frequencies was especially good (Fig. 4), the remaining differences are most likely dominated by inaccuracies in the BENM in modeling low-frequency, large-scale vibrations.

Analysis Of Allosteric Mechanisms

As demonstrated by the work of Hilser, Freire, and coworkers [19-22], it is interesting to analyze communication between remote sites in proteins in terms of conformational ensembles. In their framework, instead of describing the full atomic configurational distribution, the conformational ensemble is simplified to indicate whether residues are in a folded or unfolded state. This simplification, however, enables conformational ensembles to be calculated using methods that yield good agreement with experimentally observed hydrogen exchange protection factors for individual residues [23]. Using this framework, cooperativity effects in residue perturbations [19,22], ligand-binding [21], and intrinsic correlations in the conformational ensemble [20] have been characterized, yielding insight into allosteric effects in proteins.

Hawkins & McLeish recently have analyzed allosteric effects in terms of configurational entropy changes in Lac repressor [24]. In this study, changes in the entropy of low-frequency vibrations of the protein upon binding an allosteric regulator were calculated in a coarse-grained model. They have also analyzed vibrational free energies in coiled-coil structures such as dynein [25]. Filamentous protein structures were modeled using continuum elastic models of intertwined rods, enabling calculations of free energies due to thermal vibrations. Vibrational free energies were calculated in the presence and absence of clamping interactions at the ends of the rods, and were used to yield insight into allosteric effects.

Recently, Ming & Wall [10] used Eq. (10) to develop a general framework in which allosteric effects may be analyzed using the full configurational distribution in the harmonic approximation. In the spirit of Luque & Freire's analysis of IIA^{Glc} binding to glycerol kinase [21], changes in the configurational distribution of local regions of trypsinogen were calculated upon binding Val-Val in an allosteric site, and bovine pancreatic trypsinogen inhibitor (BPTI) in the active site. The BENM coarse-grained model was used for calculations. In binding BPTI, it was found that the local values of D_x were relatively large in the neighborhood of the BPTI-binding site. Values of D_x elsewhere on the surface were smaller, with one interesting exception: values in the Val-Val binding site, which is an allosteric site, were comparable to those in the BPTI-binding site. In addition, binding of Val-Val in the allosteric site yielded a relatively large value of D_x in the neighborhood of Ser 195, which is the key catalytic residue for trypsin and other serine proteases: the value of D_x in this neighborhood was the 40th highest of 223 residues in the crystal structure; 11th of all residues not directly interacting with the Val-Val in the model; the highest of all residues located at least as far as Ser 195 is from the Val-Val ligand; and greater than that for 20 of 60 residues located closer to the ligand. These results indicated that there is a relatively strong communication between the regulatory and active sites of trypsinogen.

THERMODYNAMIC INTERPRETATION OF D_X

Allosteric Free Energy

Let $G(x)$ be the Gibbs free energy of a protein in conformation x in solution at constant temperature and pressure. (Here we use x as a shorthand for the multidimensional configuration $\mathbf{x}$). The conformational distribution of the protein without a ligand interaction is given by

$$P(x) = Q^{-1}e^{-G(x)/k_B T},$$ (14)

where Q is the partition function. The partition function Q is related to the total free energy of the protein G as

$$G = \langle G(x) \rangle_x - TS_x,$$
$$= -k_B T \ln Q$$ (15)

25

where $\langle G(x)\rangle_x$ is the mean free energy of individual protein conformations, and $S_x = -k_B\langle P(x)\ln P(x)\rangle_x$ is the entropy of the conformational distribution. Now let $G'(x)$ be the Gibbs free energy of the protein conformation with a ligand interaction, with equations for the conformational distribution $P'(x)$ and free energy G' like those in Eqs. (14) and (15). It follows that D_x is given by

$$
\begin{aligned}
D_x &= \frac{1}{k_B T}\int dx\, P'(x)\left[-G'(x)+G(x)+k_B T\ln\frac{Q}{Q'}\right], \\
&= \frac{1}{k_B T}\left\{G'-G-\int dx\, P'(x)\left[G'(x)-G(x)\right]\right\}
\end{aligned}
\tag{16}
$$

as has been previously mentioned [11]. By the definition of the association constant $K_a(x)$ for the ligand binding to an individual protein conformation, Eq. (16) leads to

$$
k_B T D_x = \Delta G + k_B T\int dx\, P'(x)\ln K_a(x),
\tag{17}
$$

where $\Delta G = G' - G$ is the free energy difference between the protein with and without the ligand interaction. Here and elsewhere, we choose units in which the volume is unity (for other units, $K_a(x)$ should be divided by the volume to use the equations). By the top line of Eq. (15), Eq. (16) also leads to

$$
k_B T D_x = -T\Delta S_x + \int dx\,\left[P'(x)-P(x)\right]G(x),
\tag{18}
$$

where $\Delta S_x = S'_x - S_x$ is the conformational entropy difference between the protein with and without the ligand interaction. It is apparent from Equation (18) that $k_B T D_x$ is an allosteric free energy, i.e., the free energy required to change the protein conformational distribution from the equilibrium distribution without the ligand, $P(x)$, to a nonequilibrium distribution $P'(x)$ which is the same as the equilibrium distribution with the ligand bound. This statement about allosteric free energy is essentially the same as Qian's [26] association of relative entropy with the free energy of nonequilibrium fluctuations of a conformational distribution.

Experimental Measurement of Allosteric Free Energy

How might the allosteric free energy be measured experimentally? Note that Eq. (17) can be rewritten as

$$
k_B T D_x = \Delta G + k_B T\int dk_+\, P'(k_+)\ln k_+ - k_B T\int dk_-\, P'(k_-)\ln k_-,
\tag{19}
$$

where k_+ and k_- are the on and off rates for ligand binding. Although the first and third terms of Eq. (19) might be experimentally observable under biologically relevant conditions, the second term involves observing ligand binding to the protein in the ligand-bound conformational distribution, and is not generally observable. However, assuming that the protein may be frozen into a static ligand-bound conformational distribution similar to that at room temperature, analysis methods similar to those used to probe ligand recombination in myoglobin [6] might yield experimental measurements of allosteric free energy, at least in some limited cases. Such a measurement would involve flash-freezing the protein in the ligand-bound form,

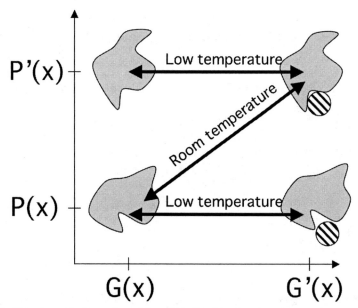

FIGURE 5. Experimental observation of ligand-binding kinetics. At room temperature, the protein conformational distribution changes depending upon whether the ligand is bound. At low temperatures, key conformational changes required for relaxation of the conformational distribution are kinetically suppressed, enabling binding kinetics to be observed for a largely "static" protein conformational distribution [6].

probing the ligand-binding kinetics at low temperature, and extrapolating the results to high-temperature (Fig. 5).

IMPLICATIONS FOR LIGAND BINDING AND ALLOSTERY

Model Of Allosteric Transitions

Equation (17) may be rewritten as

$$\Delta G = T\Delta D^a - k_B T \int dx \, P'(x)\ln K_a(x)$$
$$= -T\Delta S_x + \int dx \, [P'(x) - P(x)]G(x) - k_B T \int dx \, P'(x)\ln K_a(x). \quad (20)$$

In this section the notation is now changed such that the allosteric free energy $k_B T D_x$ is given by $T\Delta D^a$, and, as is commonly done for the entropy, k_B is absorbed into ΔD. The last term in Eq. (20) may be interpreted as the free energy change of the protein upon binding the ligand when the protein conformational distribution is fixed in its final, ligand-bound distribution. The remainder is the free energy required to change the conformational distribution from the initial, ligand-free distribution to the final, ligand-bound distribution. Equation (20) thus describes the total free energy change in a path-dependent process in which the change in the protein conformational distribution occurs independently from ligand binding (Fig. 6). First, the conformational distribution is perturbed from the equilibrium distribution, yielding a (positive) free energy change $T\Delta D^a$. Then, the ligand is bound, yielding an additional free energy change $-k_B T \int dx\ P'(x)\ln K_a(x)$ and a total free energy change ΔG.

Now consider a different path to the same state (Fig. 6): first, the ligand is added to the protein in the ligand-free distribution, changing the equilibrium distribution to $P'(x)$; then, the protein relaxes from the nonequilibrium distribution $P(x)$ to the equilibrium distribution $P'(x)$. Is the free energy change for this path the same as that for the first path? First note that by arguments similar to those leading to Eq. (17), the allosteric free energy $T\Delta D^d$ required to change the protein conformational distribution from an equilibrium ligand-bound form to a nonequilibrium ligand-free form is

$$T\Delta D^d = k_B T \int dx\ P(x)\ln\frac{P(x)}{P'(x)} ,$$
$$= -\Delta G - k_B T \int dx\ P(x)\ln K_a(x) \tag{21}$$

which leads to

$$\Delta G = -T\Delta D^d - k_B T \int dx\ P(x)\ln K_a(x)$$
$$= -T\Delta S_x + \int dx\ \left[P'(x) - P(x)\right]G'(x) - k_B T \int dx\ P(x)\ln K_a(x), \tag{22}$$

indicating that the free energies for the different paths are the same. This result is illustrated in the model in Fig. 6.

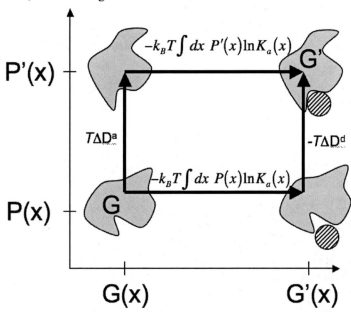

FIGURE 6. Two paths for an allosteric transition in the space of probability distributions $P(x)$ and free energy distributions $G(x)$. The protein is depicted using a large shape, and the ligand is depicted using a small circle.

Indeed, the total free energy change in this model is completely independent of the path the system takes in the space of probability distributions $P(x)$ and free energy distributions $G(x)$. For example, assume the following path for ligand binding. (1) The conformational distribution begins at the equilibrium distribution $P(x)$ when the ligand is unbound, and is changed to $\overline{P}(x)$ just before the ligand binds. (2) The free energy distribution is changed discretely from $G(x)$ to $G'(x)$ upon binding the ligand. (3) The conformational distribution relaxes from $\overline{P}(x)$ to the new equilibrium distribution $P'(x)$. By Eq. (18), step (1) entails an allosteric free energy change

$$T\Delta \overline{D} = -T(\overline{S}_x - S_x) + \int dx \left[\overline{P}(x) - P(x)\right]G(x), \qquad (23)$$

Step (2) then entails a free energy change $-\int \overline{P}(x)\ln K_a(x)$. Finally, step (3) entails an allosteric free energy change

$$-T\Delta \overline{D}' = -T(S'_x - \overline{S}_x) + \int dx \left[P'(x) - \overline{P}(x)\right]G'(x). \qquad (24)$$

Because of the relation

$$k_B T \int \overline{P}(x)\ln K_a(x) = \int \overline{P}(x)\left[G(x) - G'(x)\right], \qquad (25)$$

the total free energy change is the same as that in Eq. (20), indicating that it does not depend on the path (Fig. 7).

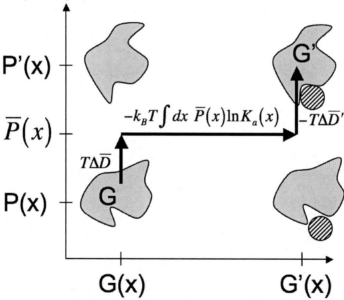

FIGURE 7. Abritrary path for allosteric ligand binding. Any paths that share the same beginning and end points in this space lead to the same change in free energy.

Thermodynamics Of A Molecular Engine

As a practical application of this framework, consider its application to analysis of the thermodynamics of a molecular engine. First, imagine a gedankenexperiment in which a single molecule may be alternately adiabatically stretched and relaxed by an external device, and whose environment may be suddenly changed to alternately cause

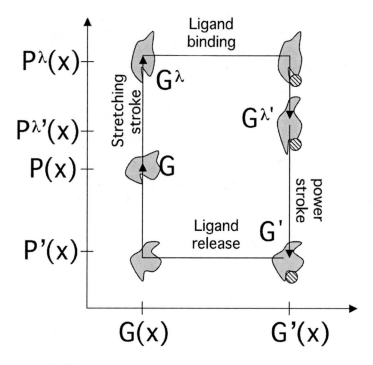

FIGURE 8. Modeling thermodynamics of a molecular engine.

complete association or dissociation of one or more ligands. This system can be thought of as an idealized real-world system such as titin being stretched using attached beads and laser traps [27]. Figure 8 illustrates a cycle in which the molecule may perform work on the external device. Suppose the molecule begins in a relaxed, ligand-free state with conformational distribution $P(x)$. During the initial stretching stroke, the conformational distribution of the molecule changes from $P(x)$ to $P^\lambda(x)$, which requires a free energy $G^\lambda - G$. Now, change the environment to cause ligand binding, leading to a new free energy $G^{\prime\lambda}$ and conformational distribution $P^{\prime\lambda}(x)$. During the subsequent power stroke, the molecule is allowed to relax, leading to a free energy change $G' - G^{\prime\lambda}$. Finally, the environment is changed to cause ligand release, returning the system to the initial state.

By the first law of thermodynamics, the usual expression for the work done on the device is

$$W = G^{\prime\lambda} - G' + G - G^\lambda = \Delta G^\lambda - \Delta G. \tag{26}$$

However, using Eq. (18) for the allosteric free energy, the work W_p done by the molecule on the device during the power stroke is

$$W_p = T\left(S'_x - S'^{\lambda}_x\right) - \int dx \, \left[P'(x) - P'^{\lambda}(x)\right]G'(x), \tag{27}$$

and the work W_s done by the device on the molecule during the stretch stroke is

$$W_s = T\left(S_x - S^{\lambda}_x\right) - \int dx \, \left[P(x) - P^{\lambda}(x)\right]G(x). \tag{28}$$

30

Using Eq. (17), it is easily shown that W as defined in Eq. (26) is equal to $W_p - W_s$. The work may thus be alternatively calculated using binding free energies or, using the present framework, allosteric free energies.

Now consider a couple of ways to make this model more realistic. First, for a real-world system, the concentration of ligand is finite. As an example, let the protein be capable of binding only a single ligand molecule, and let the equilibrium conformational distribution $P_L(x)$ be modeled as a mixture between the ligand-free distribution, $P(x)$, and the ligand-bound distribution, $P'(x)$:

$$P_L(x) = P(x)(1-f) + P'(x)f$$
$$= P(x)\big[(1-f) + K_a(x)f/V\big] \tag{29}$$

In Eq. (29), f is the fraction of time that a ligand is bound to the protein, and V is the volume of the system. At a free ligand concentration $[L]$,

$$P_L(x) = P(x)\frac{1 + K_a(x)e^{-\Delta G/k_B T}[L]}{1 + Ve^{-\Delta G/k_B T}[L]}, \tag{30}$$

which may be used to interpolate between $G(x)$ and $G'(x)$ in the above equations. Next, it is important to consider that the stretching and power strokes each take a finite time, leading to variability in the work. Assuming Jarzynski's equality [28] holds for this system, as has been experimentally demonstrated for a similar system [29], Eqs. (27) and (28) are then modified as follows

$$k_B T \ln\langle e^{W_p/k_B T}\rangle = T(S'_x - S'^\lambda_x) - \int dx\ \big[P'(x) - P'^\lambda(x)\big]G'(x), \tag{31}$$

and

$$k_B T \ln\langle e^{W_s/k_B T}\rangle = T(S_x - S^\lambda_x) - \int dx\ \big[P(x) - P^\lambda(x)\big]G(x). \tag{32}$$

Requirements For Functional Ligands

In light of Eq. (17), it is interesting to revisit the implications of the finding that values of D_x are high for interactions in small-molecule ligand-binding sites of native proteins [11] (Prediction Of Ligand-Binding Sites). On the one hand, this empirical, computational result suggests that $k_B T D_x$ is relatively high for native ligand-binding interactions. On the other hand, because the ligand binds in the binding site, ΔG should be negative and of relatively high magnitude. In considering the biological requirements of protein-ligand interactions, it is usual to emphasize the importance of ligand affinity in maximizing the total binding energy. However, by rewriting Eq. (17) as

$$k_B T \int dx\ P'(x)\ln K_a(x) = T\Delta D^a - \Delta G, \tag{33}$$

the requirements for the affinities of *functional* ligands are shown to be even more strict than is usually considered. Ligand binding to the protein in the ligand-bound conformational distribution must yield sufficient free energy not only to yield a relatively high binding energy, but also to yield a relatively high allosteric free energy.

Non-negativity of D_x also leads to some interesting relations involving binding affinities. For example, adding Eqs. (17) and (21) leads to the following inequality:

$$\int dx\ \big[P'(x) - P(x)\big]\ln K_a(x) > 0. \tag{34}$$

It also follows from Eq. (17) that, if ΔG is negative, $\int dx\, P'(x)\ln K_a(x)$ is positive, meaning that there is a net loss of free energy when the ligand binds to the protein in the ligand-bound conformational distribution; and from Eq. (21), if ΔG is positive, $\int dx\, P(x)\ln K_a(x)$ is negative, meaning that there is a net increase in free energy when the ligand binds to the protein in the ligand-free conformational distribution.

CONCLUSIONS

As reviewed above, previous studies have demonstrated the utility of the Kullback-Leibler divergence between protein conformational distributions before and after ligand binding, D_x, in analysis of ligand binding and protein vibrations [9-11]. Two especially significant outcomes of these studies are (1) a prescription for optimizing coarse-grained models of molecular vibrations; and (2) the discovery that native binding-site interactions cause a large change in protein dynamics. The former prescription is a general method that is applicable to coarse-grained modeling of proteins and other materials. The latter discovery enabled the development of a method for predicting ligand-binding sites in proteins [11]. In addition, the latter discovery suggests that sites where interactions cause a large change in protein dynamics might be well-suited to evolve as sites for controlling protein function. It also suggests that ligand-binding sites might tend to evolve at sites that are well-suited for controlling protein function.

In the rest of this paper, the thermodynamic and biophysical significance of D_x was emphasized. The quantity $k_B T D_x$ was shown to be an example of an allosteric free energy: $T\Delta D^a = k_B T D_x$ is the free energy required to change the protein conformational distribution from an equilibrium ligand-free distribution to a nonequilibrium ligand-bound distribution. Equations (17) and (18) enable the allosteric free energy to be calculated and suggest how it might be experimentally measured. They also indicate that allosteric transitions are naturally modeled in terms of motion of the system in the space of conformational distributions $P(x)$ and free-energy distributions $G(x)$.

Interestingly, the present model of allosteric transitions unifies two views of the coupling of ligand binding to conformational changes. In one view, the ligand interaction causes a perturbation in the protein conformational distribution that favors ligand binding; this view is well-aligned with Koshland's early description of induced fit, in which ligand-protein shape complementarity is only achieved when the ligand interacts with the protein [30]. In an alternative view, influence of the ligand on the protein is not significant until a perturbation in the protein conformational distribution leads to an increase in binding affinity (see, e.g., Ref. [31]). In the present context, these two views correspond to the two different pathways in Fig. 6. The first view corresponds to the path that first follows $G(x)\rightarrow G'(x)$, and then follows $P(x)\rightarrow P'(x)$. The second view corresponds to the alternative path that first follows $P(x)\rightarrow P'(x)$, and then follows $G(x)\rightarrow G'(x)$. Both paths lead to the same overall free energy change, and an argument about which of these paths might be preferred for a given system may be resolved by examining the kinetics of the different pathways and determining which is dominant in the context of the model.

As illustrated in Fig. 7, ligand-binding transitions that involve both of the above views are also possible. Determining the most probable path of *all* possibilities therefore requires consideration of not only the extreme paths that correspond to these views, but also of paths such as that illustrated in Fig. 7, and other possibilities not considered here, such as paths that involve continuous changes in the free-energy distribution $G(x)$. For example, the most probable path might first involve a fluctuation in the conformational distribution, then ligand binding, and then a subsequent relaxation to the equilibrium ligand-bound conformational distribution. Such a more general view of the coupling of ligand binding and conformational changes will be necessary to fully understand the nature of allosteric transitions in proteins.

ACKNOWLEDGMENTS

I am grateful to the Biocomputing and Physics of Complex Systems Research Institute (BIFI) for sponsoring my attendance and lecture at the Second International BIFI Congress in Zaragoza. The research described in this paper was supported by the US Department of Energy.

REFERENCES

1. D. E. Koshland, Jr., G. Nemethy, and D. Filmer, Biochemistry **5**, 365 (1966).
2. J. Monod, J. Wyman, and J. P. Changeux, J Mol Biol **12**, 88 (1965).
3. G. Weber, Biochemistry **11**, 864 (1972).
4. B. Jacrot, S. Cusack, A. J. Dianoux et al., Nature **300**, 84 (1982).
5. A. Cooper and D. T. Dryden, Eur Biophys J **11**, 103 (1984).
6. R. H. Austin, K. W. Beeson, L. Eisenstein et al., Biochemistry **14**, 5355 (1975).
7. H. Frauenfelder and P. G. Wolynes, Science **229**, 337 (1985).
8. S. Kullback and R.A. Leibler, Annals of Math Stats **22**, 79 (1951).
9. D. Ming and M. E. Wall, Proteins **59**, 697 (2005).
10. D. Ming and M. E. Wall, Phys Rev Lett **95**, 198103 (2005).
11. D. Ming and M. E. Wall, J Mol Biol (2006).
12. W. Zheng and B. R. Brooks, Biophys J **89**, 167 (2005).
13. G. Jones, P. Willett, R. C. Glen et al., J Mol Biol **267**, 727 (1997).
14. A. R. Atilgan, S. R. Durell, R. L. Jernigan et al., Biophys J **80**, 505 (2001).
15. I. Bahar, A. R. Atilgan, and B. Erman, Fold Des **2**, 173 (1997).
16. K. Hinsen, Proteins **33**, 417 (1998).
17. M. M. Tirion, Physical Review Letters **77**, 1905 (1996).
18. H. Frauenfelder and B. McMahon, Proc Natl Acad Sci U S A **95**, 4795 (1998).
19. V. J. Hilser, D. Dowdy, T. G. Oas et al., Proc Natl Acad Sci U S A **95**, 9903 (1998).
20. T. Liu, S. T. Whitten, and V. J. Hilser, Proteins **62**, 728 (2006).
21. I. Luque and E. Freire, Proteins **Suppl 4**, 63 (2000).
22. H. Pan, J. C. Lee, and V. J. Hilser, Proc Natl Acad Sci U S A **97**, 12020 (2000).
23. V. J. Hilser and E. Freire, J Mol Biol **262**, 756 (1996).
24. R. J. Hawkins and T. C. McLeish, Phys Rev Lett **93**, 098104 (2004).
25. R. J. Hawkins and T. C. B. McLeish, J. R. Soc. Interface **3**, 1742 (2006).
26. H. Qian, Phys Rev E Stat Nonlin Soft Matter Phys **63**, 042103 (2001).
27. M. S. Kellermayer, S. B. Smith, H. L. Granzier et al., Science **276**, 1112 (1997).
28. C. Jarzynski, Phys. Rev. Lett. **78**, 2690 (1997).
29. J. Liphardt, S. Dumont, S. B. Smith et al., Science **296**, 1832 (2002).
30. D. E. Koshland, Jr., J Cell Comp Physiol **54**, 245 (1959).
31. C. J. Tsai, S. Kumar, B. Ma et al., Protein Sci **8**, 1181 (1999).

Ligand Docking and Virtual Screening in Structure-based Drug Discovery

Claudio N. Cavasotto*

*MolSoft LLC, 3366 North Torrey Pines Ct. #300
La Jolla, CA 92037, United States
Email: claudio@molsoft.com

Abstract. As the number of high-resolution three-dimensional protein and nucleic acid structures continues to grow, ligand-docking—based virtual screening of chemical libraries to a receptor are playing a critical role in the drug discovery process by identifying new 'drug-candidates'. The capability to correctly predict ligand-protein interaction is fundamental to any accurate docking algorithm and the necessary starting point for any reliable virtual screening protocol. Furthermore, explicit consideration of receptor flexibility in computational ligand docking is emerging in many cases as crucial for an accurate prediction of the orientation and interactions of ligands within the binding pocket. The combination of ligand docking with a fast scoring algorithm that can account for the thermodynamics of binding, and discriminate between potential active/inactive compounds, can greatly reduced the number of compounds to be tested experimentally, while predicting a detailed structure of hits bound to the receptor useful enough to help the synthetic elaboration of leads.

Keywords: Ligand Docking, Structure-based Virtual Screening, Drug Discovery, Protein Flexibility.
PACS: 82.20.Wt, 83.10.Mj, 87.15.Aa, 87.15.Kg, 87.15.-v, 87.53.Wz

INTRODUCTION

Until not so long ago, it was believed that the identification and optimization of new leads in the drug discovery process would rely mainly on combinatorial chemistry coupled with the experimental screening of large chemical libraries against a relevant therapeutic target (High-Throughput Screening, HTS). However, it has become increasingly evident that structure-based methods should play an important role in the drug discovery process. Today, structure-based drug discovery is established as a key first step in the lengthy process of developing new drugs [1,2]. This has been fueled in recent years by several developments: novel biologically validated targets through the genomics projects; more reliable means of protein production; 3-D structural determination both by X-ray and NMR (in part due to the rapid progress of structural genomics projects, which results are becoming available to the public); advances in homology modeling methods; availability of *in silico* ligand docking and structure-based virtual screening (SBVS) programs (see Refs. [3-5] for a review on VS methods).

CP851, *From Physics to Biology; BIFI 2006 II International Congress,*
edited by J. Clemente-Gallardo, Y. Moreno, J. F. Sáenz Lorenzo, and A. Velázquez-Campoy
© 2006 American Institute of Physics 0-7354-0350-3/06/$23.00

STRUCTURE-BASED STRATEGIES IN DRUG DISCOVERY

In the early days, 3-D ligand-receptor structures (experimentally derived or homology models) have been used mainly in the drug development process to optimize ligand-receptor interactions, in an attempt to improve potency or selectivity (see Ref. [6] details a list of marketed drugs developed by using these structure-based approaches). Later, experimental and modeled protein structures were used as the starting point for *in silico* ligand docking, and screening of chemical libraries. After the recent advent of high-throughput NMR and crystallography (see Refs. [7] and [8] for review), these techniques are currently also used for fragment screening, thus enabling the construction of fragment-based targeted chemical libraries (Figure 1, and see Ref. [9] for review).

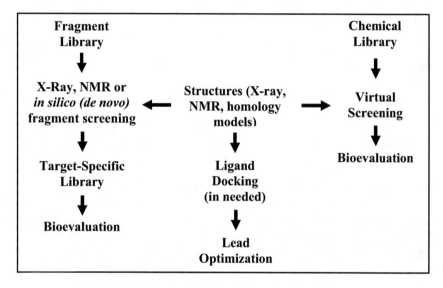

FIGURE 1. Strategies in structure-based drug discovery.

Performance of High-Throughput Screening versus Structure-Based Virtual Screening

Structural *in silico* screening methods based on docking and scoring of chemical libraries, followed by experimental evaluation are increasingly playing a critical role in the drug discovery process. Clearly, this method has many advantages compared to HTS: speed to screen a chemical library; low-cost setup to get started; low cost of failures. But, how does its performance compare to HTS?

Prof. Shoichet and his group performed a comparison of the performance of HTS vs SBVS on Protein Tyrosine Phosphatase 1B (PTP1B), a Type 2 diabetes target [10]. The hit rate for SBVS was much higher than for HTS (34.8 % vs 0.021 %), while the number of active compounds with $IC_{50} < 50$ µM was also higher for SBVS (21 vs 6). Although the screened libraries with both methods were different and the experimental conditions for HTS were not optimal, the results obtained through SBVS are more than encouraging. Moreover, active chemotypes were different using both methods.

In Table 1, a comparison performed at Merck on the tuberculosis target dihydropicolinate reductase is shown [11]. In this case, the same library was used for both methods. Remarkably, SBVS hits exhibited novel chemotypes, while those from HTS showed greater diversity.

TABLE 1. Performance of High-Throughput Screening and Structure-Based Virtual Screening on Dihydropiconlinate Reductase (Ref. [11])

Screening Method	Hit Rate $IC_{50} < 100\mu M$	IC_{50} of Best Hit (μM)
HTS	0.2 %	35
SBVS	6 %	7.2

These two examples point out the fact that HTS and SBVS do not mutually exclude, but complement each another.

Computer-based screening methods

Computational screening methods can be ligand- or structure-based. In the former case, topological, physical and/or chemical properties of known ligands are used to identify molecules that might bind to the target. The most used ligand-based methods are QSAR and pharmacophore modeling: i) the Quantitative Structure-Activity Relationships (QSAR) method represents an attempt to estimate biological activities from structural or property descriptors of compounds, based on correlations derived for existing ligands; ii) 2-D and 3-D pharmacophore models are constructed based on compounds of known biological activity. The models are then used to search virtual chemical libraries in order to find new structural classes, or as a tool for lead optimization.

When a 3-D structural representation of the receptor is available, structure-based *in silico* methods can be used. In the fragment-based *de novo* methods (see Ref. [12] for a review on this methodology and recent examples of *de novo* design), fragments are docked within the binding site and scored, and then either linked ("dock and link") or grown toward the neighboring available space ("seed and grow"). The main advantage of *de novo* methods is that they can achieve the design of novel compounds (not resembling any of those in any pre-existing chemical library). The use of small chemical groups also facilitates the exploration ligand-receptor complementarity. However, the feasibility of chemical synthesis remains a drawback for these methods. Structural virtual screening based on docking, scoring and ranking a chemical database impose a structure-based filter on compounds by selecting those with steric and electrostatic complementarity to the binding site (for review on current trends and developments in docking and SBVS see Refs. [3,4,13-18]). These compounds are further selected for experimental evaluation (for an excellent review on the computer-aided structure-based drug discovery process see Ref. [19]).

Some conceptual remarks should be made at this stage.

1. The docking process attempts the accurate prediction *in silico* of the pose of a compound within the binding site; further estimation of the binding affinity could also be attempted. A structure-based virtual screening represents the docking of a chemical

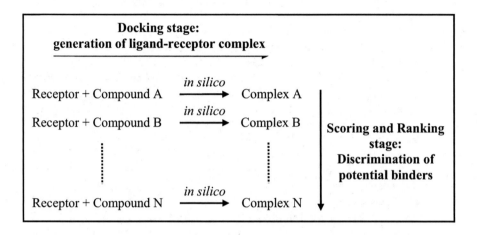

FIGURE 2. Conceptual distinction between the ligand docking, and scoring and ranking stages.

library followed by a scoring and ranking stage, in an attempt to discriminate potential binders. This is illustrated in Figure 2.

2) The goal of SBVS is to enrich the hit list with potential binders. It is not to find all of the binders in a chemical library, nor to rank them by binding affinity.

The main advantages of structure-based screening to ligand-based methods are the novelty of the hits discovered –since it is not based on pre-existing ligands-, and the possibility to determine the structure of the ligand-receptor complex, a key factor in the design of combinatorial libraries and lead optimization.

LIGAND RECEPTOR DOCKING

System Representation, Energy Function and Search Algorithms

The process of a ligand binding to a receptor takes place in solution. Thus, docking attempts to predict the structure of the ligand-receptor complex that in thermodynamical equilibrium is governed by

$$[R]_{aq} + [L]_{aq} \leftrightarrow [RL]_{aq}$$

$$\Delta G^o = -kT \ln K_a \tag{1}$$

$$K_a = \frac{[RL]_{aq}}{[R]_{aq}[L]_{aq}}$$

where R and L refers to receptor and ligand respectively, K_a is the association constant and G^o is the standard Gibbs free energy. As it has been pointed out [3], the evaluation of the complex structure ([RL]) requires calculation of the interaction and strain energy of ligand and receptor. However, evaluation of binding affinity (K_a) further requires that dissolvation effects, and conformational, rotational and translational entropy be taken into account.

There are three elements to be considered in a docking algorithm:

i)   Molecular representation: the two most used ways to represent the biomolecular system are the atomic representation and the potential energy grids. In the former case, atoms are considered explicitly and different parameters are assigned to them according to the force-field of choice (atom types, atomic charges, van der Waals parameters, etc.). The use of grids was introduced by Goodford [20], and the spirit is to represent the contributions from different terms of the potential energy on 3-D grids. This is an excellent choice on ligand-docking—based virtual screening methods due to the speed.

ii)   Docking energy function: The native conformation of the ligand should correspond to a global minimum of the energy function. The choice of the latter is obviously related to the force-field of choice, and may be supplemented by additional terms (entropy, more accurate electrostatic solvation term, etc.). Most force-fields derive their potentials and parameters from quantum mechanical calculations and/or experimental physico-chemical properties. Another possibility is to use knowledge-based parameters derived from observed inter-atomic distances.

iii)   Search algorithm: The global minimum of the energy function should be located on an usually extremely rugged energy landscape. Usual methods include Molecular Dynamics (MD), Genetic Algorithms (GA) and Monte-Carlo [21] methods.

Global Energy Minimization using the Monte Carlo Method

The main advantage of the Monte Carlo-based search algorithms is that they allow to surmount energy barriers, thus enabling a better search through the rugged energy landscape. Moreover, energy minimization in internal coordinates has been shown to converge faster than in Cartesian coordinates [22]. The main steps of a Monte Carlo-based procedure can be summarized as:

1. Random perturbation of a system variable.
2. Energy evaluation (E_{new}).
3. Is $E_{new} < E_{old}$?
   - Yes, accept the new conformation.
   - No, accept with probability $Exp(\Delta E/kT)$
4. Go back to step 1.

In dealing with proteins, performing the random step according to a predefined continue variable distribution can enhance the speed of convergence [23]. Faster convergence can also be obtained by performing a local energy minimization after the random step and before applying the Metropolis criterion [24]. The double-energy scheme, where by the minimization is performed on the differentiable variables, while the complete energy (including non-differentiable terms) is used to evaluate the metropolis criterion, has also been shown to improve convergence [25].

Examples of Ligand Docking

The binding geometry of the ligand was accurately predicted by ICM flexible docking with and without the loop regions in rhodopsin and bacteriorhodopsin. Starting from randomized side-chain conformations in the binding pocket, their correct conformation was fully restored with high accuracy (0.28Å) through the global energy optimization [26]. These binding site adjustments are critical for flexible docking of new ligands to known structures or for docking to GPCR homology models.

In the absence of experimental 3-D structures, ligand docking can be used to assess ligand-receptor binding determinants. Compound SR11179, an antagonist to the nuclear receptor human Retinoid X Receptor α (hRXRα), with an $IC_{50} \sim 1\mu M$, is very similar to SR11178, which behaves as an agonist (Figure 3). Moreover, it was found that SR11179 could be docked to hRXRα in the agonist bound conformation with no significant clashes [27].

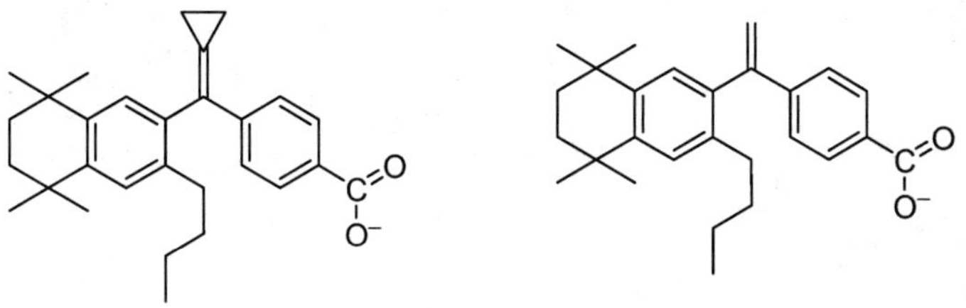

SR11179   SR11178

FIGURE 3. Chemical structure of compounds SR11178 and 11179 [27].

Full flexible docking and analysis of SR11179 into both agonist and antagonist conformations of hRXRα revealed a different binding poses, and that preferred antagonistic activity might be explain due to a better electrostatic interaction with R321. A similar hypothesis might explain the lack of human Retinoic Acid Receptor (hRARγ) agonism for SR11178 and SR11179 [27].

From a homology model of the immune modulating target Janus Kinase 3 (JAK3) protein tyrosine kinase, Adams *et al.* [28] docked the micromolar inhibitor oxindole and key ligand-receptor interactions were identified. This guided the construction of a combinatorial library that was later use for SBVS, leading to nanomolar inhibitors.

STRUCTURE-BASED VIRTUAL SCREENING

Overview of the virtual screening process

In cases where the 3-D structure of a target is available (either experimentally or modeled), SBVS is the method of choice in the early stages of the drug discovery process. Virtual screening has two different stages: the docking stage, where ligands from a chemical library are posed within the binding site; and a scoring stage, where those complexes are scored by some measure that attempts to predict the likelihood of the ligand to actually bind to the target.

A typical SBVS protocol with its pitfalls and limitations at each stage is shown in Figure 4.

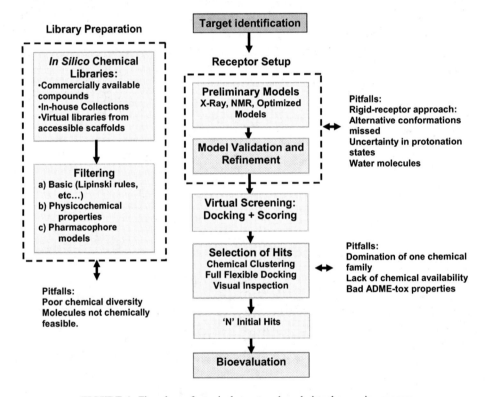

FIGURE 4. Flowchart of a typical structure-based virtual screening process.

After target identification, a 3-D representation of the target should be provided, either from experimental sources or modeled. Sometimes structures should also be refined (low-quality X-ray structure, homology model, etc.). For the sake of speed, SBVS programs use a rigid-receptor approach, thus risking to miss alternative receptor conformations (see below for a discussion on the impact of protein flexibility in docking and SBVS). Regarding the water molecules, the usual practice is to remove all bound waters, except crystallographic waters with receptor and ligand contacts.

The second component in SBVS is the choice of the chemical library. These could be commercial libraries, in-house collections, computer-generated combinatorial libraries, etc. It is advice to reduce the size of the chemical libraries by the use of proper filters. However, imposing too many filters at early stages of the process has the risk of reducing chemical diversity, though it might also reduce the number of false negatives.

DOCKING

* **Biomolecular system representation:**
  - Rigid receptor approximation (~100 faster than full atom)
  - Implicit consideration of protein flexibility.
* **Efficient search algorithm:**
  - Mone Carlo (+ minimization)
  - Genetic Algorithm
* **Docking potential** $E_{docking}$
  - Fast.
  - Discrimination among many different docking conformations of the <u>same</u> ligand.

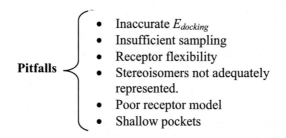

Pitfalls
- Inaccurate $E_{docking}$
- Insufficient sampling
- Receptor flexibility
- Stereoisomers not adequately represented.
- Poor receptor model
- Shallow pockets

FIGURE 5. Main steps and pitfalls of the docking stage in virtual screening.

The docking and scoring steps and pitfalls are summarized in Figures 5 and 6. It should be emphasized that $E_{docking}$, $E_{scoring}$ and $E_{bindingEnergy}$ are different functions. $E_{docking}$, the docking potential, refers to the energy of different conformations of the same ligand. In the case of true ligands, the lowest $E_{docking}$ should correspond to the native pose. $E_{scoring}$, the scoring potential, ranks the best-energy conformation of each ligand (or the few best ones), so its objective is to rank the large number of chemically diverse compounds docked to the receptor. The binding energy is too sensitive to conformational changes and thus should be carefully used as scoring function in rigid-receptor docking algorithms.

SCORING

* One or few conformations per compound are evaluated
* Scoring potential $E_{scoring}$
- Fast.
- Ranking the large number of chemically <u>diverse</u> docked compounds.

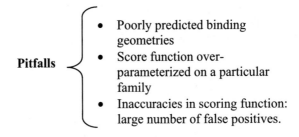

Pitfalls
- Poorly predicted binding geometries
- Score function over-parameterized on a particular family
- Inaccuracies in scoring function: large number of false positives.

FIGURE 6. Main steps and pitfalls of the scoring stage in virtual screening.

Since large chemical libraries are usually screened, scoring functions should be balanced to achieve accuracy and speed. There are force-field-based, empirical and knowledge-based scoring functions.

1) In the force-field-based methods, the score reflects the ligand-receptor interaction energy, with proper accounts of the change in the internal energy of the ligand upon binding (no such term is necessary for the receptor, since a rigid-receptor approach is usually used). Force-field energy terms should be usually supplemented with additional terms to account for solvation energy, entropy, etc.

2) In the empirical based scoring functions, $E_{scoring}$ is expressed as a linear combination of individual terms, and the parameters of the combination are adjusted to experimental data, usually the binding energy.

3) In the knowledge-based scoring, potentials are adjusted based on observed experimental structures [29,30].

Although the usual approach is to fit the empirical scoring function to experimental binding energies, an alternative approach is to fit the scoring function parameters to maximize the separation between potential binders and non-binders [31].

The top 1 to 5% of a screened database may still represent a large number of compounds to test experimentally. Thus, to reduce the number of compounds to be bioevaluated, and depending on the specific target, compounds are filtered according to different criteria, which may include, among others:

- Formation of specific ligand-receptor hydrogen bonds;
- Specific contacts between the ligand and the binding site;
- Chemical clustering for improved diversity;
- 3-D pharmacophore filtering;
- Visual inspection;
- Flexible-ligand—flexible-receptor docking for selected compounds.

Case examples of structure-based virtual screening

Some of the most popular ligand docking and virtual screening methods include AUTODOCK [32-34], DOCK [35] , GRID [20], PRODOCK [36] , EUDOC [37], FlexX [38], FLOG [39], GOLD [40], Internal Coordinate Mechanics (ICM) [41], LUDI [42], Pro_LEADS [43], QXP [44], GLIDE [45-47] and SLIDE [48]). The comparison of virtual screening methods should be performed with care, since scoring functions are parameterized for a given docking algorithm (this includes the type of representation of the biomolecular system, force-field of choice, type of conformational search, etc.) [16]. Thus, the use of scoring functions with their non-corresponding docking algorithms may lead to meaningless results.

Details of the achievements of SBVS can be found elsewhere [3,15,17-19,49]. Novel hRAR agonists in the low- and mid-nanomolar range were discovered by the use of a virtual screening algorithm followed by bioevaluation [50]. By docking to a homology model of the thyroid hormone receptor in the antagonist bound conformation, novel antagonists in the low-micromolar range were discovered and experimentally validated [51].

Very recently, the crystal structure of EGFR has been used for the first time to perform a virtual screening of a large chemical library [52]. The discovery of a C(4)-N(1) substituted pyrazolo[3,4-d]pyrimidine as a low-micromolar EGFR inhibitor with an unique binding mode opens a new avenue towards the development of novel drug scaffolds as potent and selective inhibitors of EGFR tyrosine kinase activity. It should be emphasized that this compound binds in a different way compared to the predicted mode of C(3)-C(4) substituted pyrazolo[3,4-d]pyrimidines, where N(1) and N(7) make a donor-acceptor system hydrogen bonded to the backbone [53].

PROTEIN FLEXIBILITY: IMPACT IN DOCKING AND VIRTUAL SCREENING

Statement of the problem

The mechanism by which a ligand binds to a receptor can be described with three different models. In the lock-and-key model proposed by Fischer [54] (Figure 7a), the ligand fits into the receptor like the key into a lock. Several years later, Koshland [55,56] introduced the "induced-fit" picture (Figure 7b), whereby the structures of the ligand and receptor adapt to one another upon binding. Later in 1959, Linderstrøm-Lang and Schellman reported that biomolecules can exist in many different conformations [57]. Later, Straub [58] remarked that enzymes should be described as a statistical ensemble in thermal equilibrium. Different conformations are visited according to the laws of statistical thermodynamics. This representation has been corroborated by biophysical methods. This model of preexisting conformations, together with the "induced-fit" can explain the process of ligand-protein binding. In fact, this led to the formulation of a ligand binding process in terms of linear response theory [59].

Docking to experimental structures has shown that methods which do not incorporate receptor-flexibility can usually successfully re-dock a ligand into its native structure from which it was crystallized [60]. However, when docking a molecule to a receptor which was co-crystallized with a different ligand, the methods which ignore receptor flexibility (rigid-receptor methods) may fail. Thus, high RMSD values are obtained for such cross docking. Proteins that have been crystallized with a variety of different ligands are excellent test-cases for such cross-docking experiments [60]. Examples include HIV protease [61], dihydrofolate reductase (DHFR) [62-64], HIV-RT [65] and protein kinases [66]. The implications of protein flexibility in drug discovery have been reviewed recently [67,68].

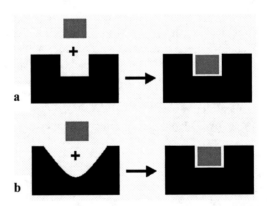

FIGURE 7. Lock-and-key (a) and induced-fit (b) models to represent protein binding. (Reprinted with permission from the Handbook of Theoretical and Computational Nanotechnology, edited by M. Rieth and W. Schommers (2006) Copyright @ American Scientific Publishers, http://www.aspbs.com)[69].

Case Example: the Impact of Protein Flexibility in Docking to Protein Kinases.

In a recent study [66], 33 crystal structures of four different protein kinases sub-families, which pockets undergo induced-fit, were used to study the impact of flexibility in rigid-receptor docking. While docking to the native conformation is always successful (usually within 1 Å RMSD), the cross-docking could be problematic. On average, only 70% of the ligands were cross-docked correctly. Moreover, it was also assessed that protein flexibility has an even stronger impact on the performance of virtual screening measured through the enrichment factor.

The enrichment factor, a measure of the relative probability of finding a ligand in the top-scoring database compared to finding a ligand at random, is given by the expression:

$$EF(n) = \frac{Hits_n}{N_n} \Big/ \frac{Hits_{total}}{N_{total}} \tag{2}$$

where n top-scoring compounds are considered.

Strategies to Incorporate Protein Flexibility in Ligand Docking and Virtual Screening.

In recent years there have been many attempts to accurately and in a realistic amount of computational time incorporate flexibility into the docking and screening algorithms. There are different levels of approximation to incorporate flexibility. The first level is the use of soft and permissive grids, where the allowed overlap between atoms constitutes a way to account for the mobility of those atoms [[70]. The second level is the use of multiple experimental structures or Receptor Ensemble Docking (RED). In this case, docking and eventually virtual screening are performed on those structures, and the results merged depending on the method of choice. The third level is the use of computer-generated multiple structures, like the IFREDA method [66] (see below). The last and most accurate level is the explicit consideration of protein flexibility. However, the latter methods are too computationally intensive to be used in the screening of large chemical libraries. The problem of flexibility and docking has been reviewed by many researchers [13,14,71,72].

In the RED methodology, the questions that still have to be answered are [13]: which should be the source of the multiple structures (X-ray, NMR, MD, MC, etc.)? How many of these structures are needed? How should these structures be used or results merged? And more importantly, how should this be incorporated in the virtual screening procedure? By using the RED strategy with two and three crystal structures of protein kinases, it was shown that significant improvements in the enrichment factor can be obtained [66]. For each compounds, the best rank –not the best score– was kept.

The ICM-Flexible REceptor Docking Algorithm (IFREDA) is especially useful in cases where only one holo (or apo) structure and known ligands are available. These ligands are used to generate alternative conformations that are later used in RED virtual screening. The improvements in enrichment factors reported are similar to those obtained using experimental structures.

Collective degrees of freedom can be also used to represent protein flexibility and study proteins dynamics [73-75], since large-scale domain movements are associated with low-energy changes and can be represented by the non-zero low-frequency normal modes. On the contrary, small movements are characterized with high-frequency modes.

A key advantage of normal mode analysis is the possibility to represent relevant receptor degrees of freedom with just a few low-frequency modes, as has been used to incorporate receptor flexibility in modeling the binding of a Hoechst 33258 analogue to DNA [76]. A low-mode conformational search procedure (LMOD) has been also developed [77-79]. But the computing time makes this latter approach unsuitable for large-scale virtual screening.

It was pointed out that the mid-scale loop displacements found in the gly-rich and C-term loops in protein kinases are not correctly represented with the first lowest-frequency modes [80,81]. By using a Cα spring-network model [82] and by introducing a *measure of relevance* of the modes on a given region of the protein, *relevant* modes are used to represent protein flexibility in protein kinases, both in cartesian space [80] or in torsion space [83]. This is very important since the representation of the conformational space grows geometrically with the number of modes. The *measure of relevance* quantifies how much each normal mode is active on a specified region, and it is seen that just few modes (< 10) are relevant in the flexible regions of cAPK binding pockets. Using cAMP-dependent Protein Kinase (cAPK, PKA) as a test case, the authors generated alternative structures by distortion along *relevant modes*.

CONCLUSIONS AND FUTURE DIRECTIONS

It is clear right now that ligand docking and structure-based virtual screening have found their place in the drug discovery process. They offer the possibility to go beyond the pool of existing active compounds and thus find novel chemotypes. When no ligands are known, they offer a rational tool for ligand discovery. Ligand docking serves also as a guide for lead optimization in cases where no experimental structure of the complex is available.

In the algorithmic field, improvements in scoring functions are still needed. For example, a better balance between polar and non-polar interactions, and improvements in the solvation and entropy term. The systematic incorporation of protein flexibility in structure-based virtual screening is a real challenge, though many groups are actively working on this regard.

ACKNOWLEDGMENTS

Discussions with Prof. Ruben Abagyan and Dr. Andrew Orry are greatly acknowleged.

REFERENCES

1. L. M. Amzel, Curr Opin Biotechnol **9**, 366-9 (1998).
2. G. Klebe, J Mol Med **78**, 269-281 (2000).
3. D. B. Kitchen, H. Decornez, J. R. Furr, and J. Bajorath, Nat Rev Drug Discov **3**, 935-49 (2004).
4. E. M. Krovat, T. Steindl, and T. Langer, J. Comput.-Aided Drug Des. **1**, 93-102 (2005).
5. R. D. Taylor, P. J. Jewsbury, and J. W. Essex, J Comput Aided Mol Des **16**, 151-66 (2002).
6. M. Congreve, C. W. Murray, and T. L. Blundell, Drug Discov Today **10**, 895-907 (2005).
7. H. O. Villar, J. Yan, and M. R. Hansen, Curr Opin Chem Biol **8**, 387-91 (2004).
8. T. L. Blundell and S. Patel, Curr Opin Pharmacol **4**, 490-6 (2004).
9. A. J. W. Orry, R. A. Abagyan, and C. N. Cavasotto, Drug Discov. Today **11**, 261-266 (2006).
10. T. N. Doman, S. L. McGovern, B. J. Witherbee, T. P. Kasten, R. Kurumbail, W. C. Stallings, D. T. Connolly, and B. K. Shoichet, J. Med. Chem. **45**, 2213-21 (2002).
11. A. M. Paiva, D. E. Vanderwall, J. S. Blanchard, J. W. Kozarich, J. M. Williamson, and T. M. Kelly, Biochim Biophys Acta **1545**, 67-77 (2001).
12. G. Schneider and U. Fechner, Nat Rev Drug Discov **4**, 649-63 (2005).
13. H. A. Carlson, Curr Pharm Des **8**, 1571-8 (2002).
14. C. N. Cavasotto, A. J. W. Orry, and R. Abagyan, Curr. Comput. Aided Drug Des. **1**, 423-440 (2005).
15. B. K. Shoichet, Nature **432**, 862-5 (2004).
16. J. C. Cole, C. W. Murray, J. W. Nissink, R. D. Taylor, and R. Taylor, Proteins **60**, 325-32 (2005).
17. R. Abagyan and M. Totrov, Curr Opin Chem Biol **5**, 375-82 (2001).
18. J. C. Alvarez, Curr Opin Chem Biol **8**, 365-70 (2004).
19. A. C. Anderson, Chem Biol **10**, 787-97 (2003).
20. P. J. Goodford, J Med Chem **28**, 849-57 (1985).
21. N. A. Metropolis, A. W. Rosenbluth, N. M. Rosenbluth, A. H. Teller, and E. Teller, J. Chem. Phys. **21**, 1087-92 (1953).
22. R. Abagyan, in *Computer simulation of biomolecular systems*; *Vol. 3*, edited by W. F. van Gunsteren, P. K. Weiner, and A. J. Wilkinson (Kluwer/Escom, Dordrecth, The Netherlands, 1997), p. 363-394.
23. R. Abagyan and M. Totrov, J Mol Biol **235**, 983-1002 (1994).
24. Z. Li and H. A. Scheraga, Proc Natl Acad Sci U S A **84**, 6611-5 (1987).
25. R. Abagyan, M. Totrov, and D. Kuznetsov, J. Comput. Chem. **15**, 488-506 (1994).
26. C. N. Cavasotto, A. J. Orry, and R. A. Abagyan, Proteins **51**, 423-433 (2003).
27. C. N. Cavasotto, G. Liu, S. Y. James, P. D. Hobbs, V. J. Peterson, A. A. Bhattacharya, S. K. Kolluri, X. K. Zhang, M. Leid, R. Abagyan, R. C. Liddington, and M. I. Dawson, J. Med. Chem. **47**, 4360-4372 (2004).
28. C. Adams, D. J. Aldous, S. Amendola, P. Bamborough, C. Bright, S. Crowe, P. Eastwood, G. Fenton, M. Foster, T. K. Harrison, S. King, J. Lai, C. Lawrence, J. P. Letallec, C. McCarthy, N. Moorcroft, K. Page, J. Redford, S. Sadiq, K. Smith, J. E. Souness, S. Thurairatnam, M. Vine, and B. Wyman, Bioorg Med Chem Lett **13**, 3105-10 (2003).
29. H. Gohlke, M. Hendlich, and G. Klebe, Journal of Molecular Biology **295**, 337-356 (2000).
30. I. Muegge and Y. C. Martin, Journal of Medicinal Chemistry **42**, 791-804 (1999).
31. M. Totrov and R. Abagyan, in *Derivation of sensitive discrimination potential for virtual ligand screening*, Lyon, France, 1999 (Associaton for Computer Machinery, New York), p. 37-38.
32. D. S. Goodsell, G. M. Morris, and A. J. Olson, J Mol Recognit **9**, 1-5 (1996).
33. G. M. Morris, D. S. Goodsell, R. S. Halliday, R. Huey, W. E. Hart, R. K. Belew, and A. J. Olson, J. Comp. Chem. **19**, 1639-1662 (1998).
34. G. M. Morris, D. S. Goodsell, R. Huey, and A. J. Olson, J Comput Aided Mol Des **10**, 293-304 (1996).
35. T. J. A. Ewing and I. D. Kuntz, Journal of Computational Chemistry **18**, 1175-1189 (1997).
36. J. Y. Trosset and H. A. Scheraga, Journal of Computational Chemistry **20**, 412-427 (1999).
37. E. Perola, K. Xu, T. M. Kollmeyer, S. H. Kaufmann, F. G. Prendergast, and Y. P. Pang, Journal of Medicinal Chemistry **43**, 401-408 (2000).
38. M. Rarey, B. Kramer, T. Lengauer, and G. Klebe, Journal of Molecular Biology **261**, 470-489 (1996).

39. M. D. Miller, S. K. Kearsley, D. J. Underwood, and R. P. Sheridan, Journal of Computer-Aided Molecular Design **8**, 153-174 (1994).
40. G. Jones, P. Willett, R. C. Glen, A. R. Leach, and R. Taylor, J Mol Biol **267**, 727-48 (1997).
41. R. Abagyan, M. Totrov, and D. Kuznetsov, Journal of Computational Chemistry **15**, 488-506 (1994).
42. H. J. Bohm, Journal of Computer-Aided Molecular Design **6**, 61-78 (1992).
43. C. A. Baxter, C. W. Murray, D. E. Clark, D. R. Westhead, and M. D. Eldridge, Abstracts of Papers of the American Chemical Society **216**, U693-U693 (1998).
44. C. McMartin and R. S. Bohacek, Journal of Computer-Aided Molecular Design **11**, 333-344 (1997).
45. T. Halgren, Abstracts of Papers of the American Chemical Society **220**, U168-U168 (2000).
46. R. Friesner, T. A. Halgren, P. Shenkin, R. B. Murphy, J. Banks, H. Beard, J. Klicic, D. Mainz, J. Perry, and T. F. Hendrickson, Abstracts of Papers of the American Chemical Society **223**, U465-U465 (2002).
47. T. A. Halgren, R. B. Murphy, J. Banks, D. Mainz, J. Klicic, J. K. Perty, and R. A. Friesner, Abstracts of Papers of the American Chemical Society **224**, U345-U345 (2002).
48. V. Schnecke, C. A. Swanson, E. D. Getzoff, J. A. Tainer, and L. A. Kuhn, Proteins-Structure Function and Genetics **33**, 74-87 (1998).
49. B. K. Shoichet, S. L. McGovern, B. Wei, and J. J. Irwin, Curr Opin Chem Biol **6**, 439-46 (2002).
50. M. Schapira, B. M. Raaka, H. H. Samuels, and R. Abagyan, BMC Structural Biology **1**, 1 (2001).
51. M. Schapira, B. M. Raaka, S. Das, L. Fan, M. Totrov, Z. Zhou, S. R. Wilson, R. Abagyan, and H. H. Samuels, Proc Natl Acad Sci U S A **100**, 7354-9 (2003).
52. C. N. Cavasotto, M. A. Ortiz, R. A. Abagyan, and F. J. Piedrafita, Bioorg Med Chem Lett **16**, 1969-74 (2006).
53. P. Traxler, G. Bold, J. Frei, M. Lang, N. Lydon, H. Mett, E. Buchdunger, T. Meyer, M. Mueller, and P. Furet, J Med Chem **40**, 3601-16 (1997).
54. E. Fischer, Ber. Dtsch. Chem. Ges, 2985 (1894).
55. D. E. Koshland, Jr., Angewandte Chemie-International Edition **33**, 2375-2378 (1995).
56. D. E. Koshland, Jr., Proc Natl Acad Sci U S A **44**, 98-123 (1958).
57. K. U. Linderstrøm-Lang and J. A. Schellman, in *The Enzymes*; *Vol. 1*, edited by L. Boyer and Myrbäk (Academic Press, New York, 1959), p. 443-510.
58. F. B. Straub, Adv Enzymol Relat Areas Mol Biol **26**, 89-114 (1964).
59. M. Ikeguchi, J. Ueno, M. Sato, and A. Kidera, Phys. Rev. Lett. **94**, 078102 (2005).
60. B. Kramer, M. Rarey, and T. Lengauer, Proteins **37**, 228-41 (1999).
61. D. Bouzida, P. A. Rejto, S. Arthurs, A. B. Colson, S. T. Freer, D. K. Gehlhaar, V. Larson, B. A. Luty, P. W. Rose, and G. M. Verkhivker, International Journal of Quantum Chemistry **72**, 73-84 (1999).
62. B. I. Schweitzer, A. P. Dicker, and J. R. Bertino, Faseb Journal **4**, 2441-2452 (1990).
63. V. Cody, N. Galitsky, D. Rak, J. R. Luft, W. Pangborn, and S. F. Queener, Biochemistry **38**, 4303-4312 (1999).
64. J. Feeney, Angewandte Chemie-International Edition **39**, 290-312 (2000).
65. F. Daeyaert, M. de Jonge, J. Heeres, L. Koymans, P. Lewi, M. H. Vinkers, and P. A. Janssen, Proteins **54**, 526-33 (2004).
66. C. N. Cavasotto and R. A. Abagyan, J Mol Biol **337**, 209-225 (2004).
67. A. M. Davis, S. J. Teague, and G. J. Kleywegt, Angew Chem Int Ed Engl **42**, 2718-36 (2003).
68. S. J. Teague, Nat Rev Drug Discov **2**, 527-41 (2003).
69. C. N. Cavasotto, A. J. W. Orry, and R. Abagyan, in *Handbook of Theoretical and Computational Nanotechnology*, edited by M. Rieth and W. Schommers (American Scientific Publishers, in press, 2006).
70. F. Jiang and S. H. Kim, J Mol Biol **219**, 79-102 (1991).
71. H. A. Carlson and J. A. McCammon, Mol Pharmacol **57**, 213-8 (2000).
72. M. L. Teodoro and L. E. Kavraki, Curr Pharm Des **9**, 1635-48 (2003).
73. D. A. Case, Curr. Opinion Struct. Biol. **4**, 285-290 (1994).
74. S. Hayward and N. Go, Annu. Rev. Phys. Chem. **46**, 223-250 (1995).
75. T. Noguti and N. Go, Nature **296**, 776-778 (1982).
76. M. Zacharias and H. Sklenar, J. Comput. Chem. **20**, 287-300 (1999).

77. I. Kolossváry and W. C. Guida, J Am Chem Soc **118**, 5011-5019 (1996).
78. I. Kolossváry and W. C. Guida, J Comput Chem, 1671-1684 (1999).
79. G. M. Keserü and I. Kolossváry, J Am Chem Soc **123**, 12708-9 (2001).
80. C. N. Cavasotto, J. A. Kovacs, and R. A. Abagyan, J. Am. Chem. Soc. **127**, 9632-9640 (2005).
81. A. May and M. Zacharias, Biochim Biophys Acta **1754,** 225-31 (2005).
82. M. M. Tirion, Phys. Rev. Lett. **77,** 1905-1908 (1996).
83. J. A. Kovacs, C. N. Cavasotto, and R. A. Abagyan, J. Comp. Theor. Nanosci. **2,** 354-361 (2005).

The mechanical opening of DNA and the sequence content

S. Cocco* and R. Monasson†

*CNRS-Laboratoire de Physique Statistique de l'ENS, 24 rue Lhomond, 75005 Paris, France
†CNRS-Laboratoire de Physique Théorique de l'ENS, 24 rue Lhomond, 75005 Paris, France

Abstract.
Separation of the strands of the DNA molecule play a fundamental role in biology. We here review different experimental and theoretical approaches to DNA opening in vitro. We then analyze the relationship between the sequence content and the opening signal. We focus on a theoretical study of the performances of Bayesian inference to predict the sequence of DNA molecules from fixed-force unzipping experiments.

Keywords: DNA Sequence, Mechanical Unzipping, Bayesian Inference
PACS: 87.14.Gg, 02.50.Ey, 02.30.Zz

INTRODUCTION

In all living cells long DNA molecules carry genetic information in their base-sequence. This information is read and processed to make proteins during transcription, to duplicate DNA during replication before cellular division and to repair DNA when it is damaged [1, 2, 3].

The DNA double helix is made of two polymeric chains, wound around one another to form a regular, right-handed helix. Each chain is a series of chemical units called nucleotides, joined together by single covalent bonds. The four types of nucleotides - containing the 'bases' adenine (A), thymine (T), guanine (G) and cytosine (C) - can be chained together in any order; the genetic information is carried precisely by the sequence of bases on a chain. The double-helix structure is only stable if the two chains carry complementary sequences, meaning juxtaposed A-T and G-C base pairs. The double helices found in cells obey this constraint (apart from damaged nucleotide) and thus can be reconstructed from only one strand, simply by synthesis of a complementary strand.

The two oppositely directed chains are bound together by hydrogen bonds and other physical interactions which generate only about $k_B T$ of net cohesive free energy per base pair under conditions found in the cell (aqueous solution at room temperature with about 0.1 M salt and pH near 7.5). This means that double helices of more than about 30 base pairs in length are stable (they require thermal fluctuations $> 30 k_B T$ to fall apart, which are rare enough to be ignored). At the same time, the two strands can easily taken apart without breaking the covalent backbone bonds. This strand separations takes place, indeed, if the solution condition are changed by increasing the temperature, increasing the pH, or decreasing the ionic condition [2].

CP851, *From Physics to Biology; BIFI 2006 II International Congress,*
edited by J. Clemente-Gallardo, Y. Moreno, J. F. Sáenz Lorenzo, and A. Velázquez-Campoy
© 2006 American Institute of Physics 0-7354-0350-3/06/$23.00

The two DNA strands can also be separated mechanically. Micromechanical experiments allows one to gradually separate, or unzip, the two strands of long DNAs by the application of a force [4, 5, 6, 7, 8]. Finally strand separation can be also performed by enzymes, as is done in the cell; single molecule experiments follow a single enzyme working on DNA [9, 10, 11, 12, 13, 14].

Hereafter we will review some DNA strands separation or hybridization by in vitro experiments, from denaturation experiments (section) to single molecule experiments (section). Among single molecule experiments [11] we will review the detection of single strand DNA hybridization [16], the DNA strand separation under translocation through a nano-pore [17, 18], the observation of the sequence-dependent activity of an exo-nuclease [9, 10] and a DNA polymerase [11, 12, 13], and the unzipping of DNA under a mechanical action at a constant velocity [4, 5, 7] or force [6, 8].

We will in particularly focus on how information on the DNA sequence can be extracted from experimental data. The work to separate the DNA strands depends indeed on the sequence, not only because the base pairs GC are most stable than the base pairs AT but also because there are base-dependent stacking interactions between neighboring base pairs in the double helical structure. The knowledge of the exact base content of DNA's sequence is of huge importance from both biological and medical points of view. Biochemical sequencing methods, described in section , are very powerful and have been recently massively used for sequencing whole genomes, in particular the human one. Moreover looking for alternative way of sequencing DNA is an active field of research. In this regards DNA strand separation experiments on single molecules of DNA could be interesting. So far the single molecule experiments have however not been combined with a concrete method for data analysis capable of extracting precise information on the sequence of DNA molecules. In section we will review some theoretical studies on the prediction of a sequence from fixed force unzipping experiments.

STRAND SEPARATION OR HYBRIDIZATION

Thermal Denaturation of a DNA solution

The melting transition is, in classical biochemistry experiments, observed by measuring the optical properties of a solution containing DNA molecules, as the ultraviolet absorbance or the chromatic dichroism which are sensitive to the amount of double stranded regions in the molecule [2]. Alternatively DNA can be melted in calorimeters. In this set-up the melting transition is detected from the peaks in the heat capacity.

When raising the temperature the change in optical properties or in the heat capacity occur over a narrow temperature range (a few tenth of degrees), revealing that denaturation is a cooperative process. The melting temperature, defined as the temperature value at half the transition is then quite well defined. The melting temperature ranges between 60^oC and 110^oC [2]; it depends on the sequence, on the length of the molecule and the chemical properties (pH, salt concentration) of the solution.

For a repeated sequence the melting transition is very sharp. The base pairing free energy can be directly derived from the melting temperature. The pairing energies of the 16 possible combinations of two consecutive base pairs (note that due to complementarity

TABLE 1. Pairing free energies $g_0(b, b')$ from the Mfold server [19] and Santa Lucia data [20] at temperature $T = 25^{o}C$, 150 mM NaCl and neutral pH. The columns refers to b', the lines to b in the $5' \rightarrow 3'$ direction along the molecule. The pairing free energies are given by the hydrogen bonds between the two bases of a same pair and stacking interactions between neighboring bases, see section for more detail.

	A	T	C	G
A	1.78	1.55	2.52	2.22
T	1.06	1.78	2.28	2.54
C	2.54	2.22	3.14	3.85
G	2.28	2.52	3.90	3.14

between strands there are only 10 different values) have been indeed precisely quantified from the experimental melting temperature, obtained in calorimeters, of artificial sequences made by the repetition of the two base pairs motive [19, 20] (see Table 1) .

In the denaturation of heterogeneous sequences the melting temperature gives only information on the average sequence. The derivative of the absorbance melting curve as well as the peaks in the heat capacity give a more detailed information on the sequence. It reveals specific regions in the DNA that melt cooperatively over a narrow temperature range. AT-rich regions melt first, at lower temperature, and the GC-rich regions melt at higher temperature (see sketch in Fig. 1). Analysis of differential melting data can reveal specific characteristics of the sequence such as the presence of regions of special stability (rich in GC or in AT) or with special sensitivity to specific ions, or also the presence of mismatches. Melting data can thus be used as a rapid method to identify homology or differences among DNAs. However this method does not allow one to localize the identified regions into the DNA sequence.

Single molecule experiments

Strand separation and hybridization has recently been observed by mechanical micromanipulation of single DNA molecules [4, 5, 6, 7, 8, 9, 10, 11, 12, 13, 14]. By micromanipulation experiments a single molecule can be stretched by a force while measuring its extension. Experiments and theory have allowed to characterize the elasticity of a single strand DNA molecule (ssDNA) as well as the one of double strand DNA molecule (dsDNA) [21, 22, 23, 24, 25]. The force extension curves at room temperature and standard ionic conditions are shown in Figure 2. At small forces the ssDNA and dsDNA look like random coils, but the ssDNA is more flexible than the dsDNA. Up to forces of 5 pN the ssDNA is indeed shorter than the dsDNA . On the contrary at forces larger than 5 pN

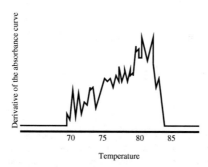

FIGURE 1. Schematic representation of the differential melting curve for the λ phage sequence (see [2] for the exact plot)

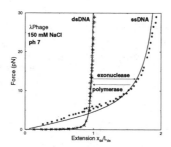

FIGURE 2. Stretching curve of the λ phage DNA molecule, in its double strand conformation (dsDNA) and single strand conformation (ssDNA). dsDNA curve: x symbols are experimental data from [23], + symbols are data from [22], straight line is theory from [24]. ssDNA curve: data (filled circles) and theory (straight line) from [21]. L_{ds} is the crystallographic length of the ds-λ DNA. The ssDNA is shorter than the dsDNA for forces smaller than $f \simeq 5$ pN and longer for larger forces. The enzymes exonuclease convert dsDNA into ssDNA while the polymerases make the opposite.

both the ssDNA and the dsDNA are essentially straight up by the force; the length of the molecule is therefore the number of nucleotides times the distance between consecutive bases, of about 7 Å for the ssDNA backbone distance and 3.4 Å for the dsDNA axial distance. Therefore, in the range of forces of 5-60 pN, ssDNA has about twice the length of dsDNA.

The detection of a single DNA hybridization based on the different lengths of ssDNA and dsDNA has been experimentally performed by Singh-Zocchi and collaborators [16]. A small ssDNA fragment is attached between a glass slides and a bead that can be pulled by a force, while the complementary strand is put in the surrounding solution (Fig. 3). When the two strand hybridize in dsDNA the length of the molecule change. The change in length is detected by a fluorescence resonance energy transfer method. A second method that allows one to measure in a similar way the opening time of small DNAs is nanopore unzipping [17, 18]. In this experiment (Fig. 5) a DNA hairpin with a ssDNA tail at one extremity is first catched in a micropore by a voltage that is applied on the opposite extremity of the channel. Because the nanopore is very thin, the DNA hairpin

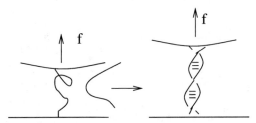

FIGURE 3. Single molecule hybridization of a short DNA, attached from one hand to a bead and by the other hand to a surface stretched by a force $f < 5$ pN (at which the ssDNA is shorter than dsDNA as shown in Fig.2). The conformational change due to the hybridization with a free oligomers is optically detected [16].

has to open to pass through the nanopore. The presence of the DNA in the channel can be detected by the reduction of the current passing through the channel. The time that the DNA hairpin takes to open can be therefore measured. These experiments detect the overall hybridization or opening times which depend on the sequence, but, similarly to the thermal denaturation, allow one only to have a global information on the sequence.

Let us now describe method that give local information on the sequence content. A method recently developed is based on the single molecule detection of the polymerization or depolarization of a single DNA by enzymes. Enzymes called DNA polymerase synthesize dsDNA from ssDNA and a solution containing free nucleotides. Enzymes called exonucleases make the opposite: they remove one single strand from the double helix by cleaving one nucleotide after the other. Thanks to single molecule experiments, a single ssDNA or dsDNA molecule can be stretched by a micromanipulator and the activity of such enzymes can be monitored by the change in molecular length consequent to their action [9, 10, 11, 13, 12]. Interestingly, the experiments have shown that the rate of depolymerization by an exonuclease depends on the sequence. This rate dependence on the sequence demonstrate that enzymes feel the (different) binding energies of nucleotides while processing the sequence [9]. Moreover long pauses occur in correspondence with sequence-specific motif *eg.* GGCGA on the λ-phage sequence. These experiments therefore give the information on the sequence content and position on it. These methods are however, at first sight, quite involved to use as sequencing methods because enzyme are complicated and stochastic machines which for example undergo conformational changes during their activity. To obtain information on the sequence the sequence dependent fluctuations in the processing rate have to be separated from fluctuations independent on the sequence and attributed to conformational changes [9]. In the following we focus on the unzipping of the DNA molecule by a force and not an enzyme.

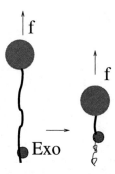

FIGURE 4. The activity of an exonuclease on a single stretched DNA can be detected by the shortening of the molecule [9, 10] stretched by a force $f < 5$ pN.

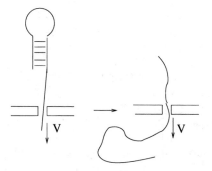

FIGURE 5. Nanopore unzipping: a small dsDNA hairpin to which is attached a ssDNA link is "fished" into a channel in a membrane by applying a Voltage through the channel. The dsDNA cannot pass through the pore and therefore it goes through only after having been opened by thermal fluctuations [17, 18] .

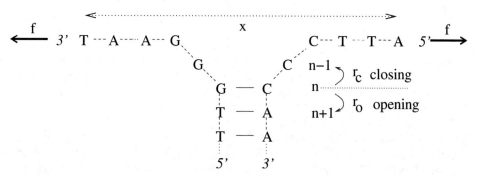

FIGURE 6. Sketch of the fixed force opening. In the opening model the fork in position n (number of open bases) move forward or backward with rates (probabilities per time) dependent on the free energy potential $G(n, f)$, see text and Figure 7.

DNA SEQUENCING METHODS

Natural DNA are usually too long to be sequenced directly. They are first fragmented into pieces of maximal length of about 1000 nucleotides which are then sequenced. The position of fragments within the large DNA are then reconstructed by mapping the regions with overlapping sequences between fragments. To be sequenced a DNA fragment is first cloned into many copies, usually by a PCR reaction. The DNA sequence of the fragments are then determined by chemical or enzymatic methods. The Maxam-Gilbert chemical method depends on base specific chemical reactions to break polynucleotide chains specifically at A,G, C or T. A ssDNA fragment is first labeled at either its 5'-or-3' end, then treated with a base specific reagent (*e.g.* specific for G), so that less than 1% of the G's react. Further chemical treatment causes strand breakage at each modified G. The breakage produces a set of end-labeled fragments whose lengths reveals the positions of G's within the sequence, the length can be determined with very high precision by gel electrophoresis. Many other fragments result but only the end-labeled ones are analyzed. Use of four base specific reactions in separate experiments provides the position of each base relative to the end label, *i.e.* the sequence. The principle of the enzymatic Sanger method is the same as that of Maxam and Gilbert: the length of an end-labeled oligonucleotide that terminates at one type of base reveals the position of that base. In the Sanger methods a complementary strands is synthesized. A small fraction of a single type of didexyonucleoside triphosphate *e.g.* didexyguanosine 5' triphosphate is used to cause chain termination during synthesis. Gel electrophoresis is then used to determine the chain lengths of the set of end labeled oligonucleotides.

These techniques are very powerful and reliable, with a 99.9% fraction of correctly predicted bases. However they are costly in terms of time and workforce. In addition some regions in the genome may be very difficult to clone, and obtaining information on their sequence is uneasy.

UNZIPPING:EXPERIMENTS AND THEORY

In 1997, U. Bockelmann and F. Heslot demonstrated a new way to obtain information on the DNA base sequence, based on single molecule micromanipulation and force measurements [4]. The idea consists in unzipping the molecule by pulling apart the two complementary strands of the double helix (Fig. 6). If one opens the double helix by separating the extremities of the two single strands at constant velocity for instance, one can measure the force as a function of time, *i.e.* as a function of the number of opened base pairs. If one pulls slowly the unzipping force is of about 15 pN. The positive or negative excursions of the force around this average value reproduce the base content of the sequence, rich or poor in GC base pairs, respectively [4, 5]. The binding energy of the GC pair is higher than the one of the AT pair, hence a higher force is needed to break a GC pair.

With a more recent experimental setup Bockelmann and Heslot were able to observe the GC content of the sequence of long molecules (about 10,000 base pairs) with a sensitivity of about 10 DNA base pairs. Remarkably, a comparison between measured and calculated force signals (using the known DNA sequence) indicates that the signal could be sensitive to certain point mutations [5].

DNA unzipping appears as a mechanical way to read the sequence. Moreover the possibility to change the distance between the ssDNA extremities of the molecule, allows one to read the bases one after the other. Two information are collected together during unzipping: the distance between the ssDNA extremities of the molecule (Fig. 6) that is related to the position of the opening base and the force related to the type of the opening base.

Several limitations however do not allow one to directly read the sequence and lead to the feeling that sequencing by unzipping is not a achievable tool [26]. The first problem is that unzipping is stochastic: the work to open a base pair is only few $k_B T$ so a base pair can open and close by thermal fluctuations. Two repetitions of an unzipping experiment give different data. The second problem is that unzipping traces do not give access directly to the position of the opening fork $n(t)$, but to the distance $x(t)$ between the extremities of the molecule (Fig. 6). The unzipped single strands that link the extremities of the molecule and the opening fork are not rigid, but fluctuate. The value of $x(t)$ for a given $n(t)$ is therefore stochastic due to the thermal fluctuations of the unzipped single strands. Finally there are additional sources of noises from the experimental apparatus e.g. drift of the apparatus, and temporal and spatial resolutions.

Yet unzipping is a potentially powerful tool, since long molecules can be opened by unzipping in a fraction of seconds, and one single molecule can be opened and closed several times, to collect more opening signal. Our aim is to theoretically understand if extracting information on the sequence is possible despite this different sources of noise. We would like to design optimal protocols and methods of data analysis capable of providing us with the most accurate information possible about the sequence.

In the following we focus on computer-generated data, for which the dynamical model and the source of noise are perfectly well defined.

To start with we have considered the fixed force unzipping [6, 8], in which a force is kept constant at the extremities of the molecule while the distance between them is measured.

As a first step we have focus only on the thermal movement of the fork position $n(t)$. In Section we introduce a simple model for the unzipping at constant force able to reproduce the experimental data. In Section we discuss the prediction of the sequence from unzipping data generated from this model by a Monte Carlo simulation.

Finally we discuss how to tackle the more realistic case with several sources of noise.

A model for fixed force unzipping

For artificial repeated sequences, the unzipping at fixed force f is easy to describe: the free energy difference $\Delta g(f)$ between two open bases and the close base pair is simply the difference between the elastic free energy $w_{ss}(f)$ of the two unzipped DNA bases submitted to a force f, and the binding energy g_0 of the base pair,

$$\Delta g(f) = 2w_{ss}(f) - g_0. \tag{1}$$

The critical force, at which the molecule unzip is therefore the value f^* such that $\Delta g(f^*) = 0$. The free energy $w_{ss}(f)$ is well known from ssDNA stretching experiments [21, 25]; it is the integral of the force-extension curve of Fig.2. The knowledge of f^* allows us to predict the binding free energy g_0 for repeated sequences [15]. Unzipping of repeated sequence has been performed by Rief *et al.* They have found $f^* = 20 \pm 3$ pN for poly(dG-dC) and $f^* = 9 \pm 3$ pN for poly(dA-dT). These forces give, $g_0 = 3.5 k_B T$, $g_0 = 1.1 k_b T$ respectively for the poly(dG-dC) and poly(dA-dT) sequence [25], in good agreement for the data of Table 1 obtained from thermal denaturation. For a heterogeneous sequence $\mathscr{S} = \{b_1 \ldots b_N\}$, the free energy to unzip base pair i,

$$\Delta g(f, i) = 2 w_{ss}(f) - g_0(b_i, b_{i+1}) \tag{2}$$

depends on the base and its neighbors due to stacking effects. The difference in free energy of the molecule when its first n bases are unzipped and when it is fully zipped is then

$$G(n, f) = \sum_{i=1, n} \Delta g(f, i). \tag{3}$$

The free energy landscape $G(n, f)$ for the first 450 bases of λ phage DNA is plotted in Fig. 7 C, at a force of 16.4 pN; the overall slope is negative, meaning that at this force the molecule opens. The free energy landscape is characterized by sequence-dependent barriers and minima between them. A simple modeling able to reproduce the unzipping dynamics (Fig. 2) describes the dynamic of the border (hereafter referred to as the fork) between the open and closed regions of the DNA molecule [5, 25, 27, 28, 29, 30, 31]. The motion of the fork, whose position corresponds to the number $n(t)$ of open base pairs at time t, is modeled as a one-dimensional random walk in the potential $G(n, f)$. Fig. 7 A shows pseudo unzipping data (in silico experiments) obtained from Monte Carlo simulation. A time-trace is the sequence $\mathscr{N} = \{n_i\}$ of the fork positions at discrete times $t = i \times \Delta$, where Δ is the delay between two measures (inverse of the temporal bandwidth). Fig. 7B and Fig. 7C show that the plateaus in the opening signal are in correspondence with the deepest minima of the free energy $G(n, f)$. Our model implicitly defines the probability $\mathscr{P}(\mathscr{N}|S)$ to measure a time-trace $\mathscr{N}$ given the sequence S of the molecule.

Data treatment in fixed force-unzipping: from the signal to the sequence

Up to now we have considered the direct problem: from a given sequence $\mathscr{S} = \{b_i\}$ *e.g.* the λ-phage we have defined a model that gives un unzipping trace reproducing the experimental data. As we have seen the unzipping trace is just a unidimensional random walk in the free energy landscape $G(n, f)$ (function of the sequence $\mathscr{S}$)

The inverse problem can be stated in general terms as follows: from the trace $\mathscr{N} = \{n(t)\}$ of a discrete random walk is it possible to reconstruct the free energy landscape $G(n, f)$? This question translates, in our case, as follows: from the temporal trace $\mathscr{N}$ of the opening force can we deduce the sequence $\mathscr{S}$? We have studied this inverse problem

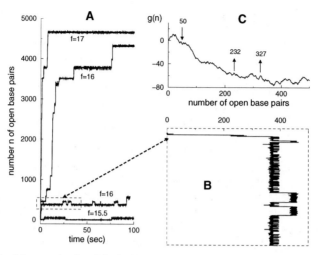

FIGURE 7. Fixed force unzipping of the λ phage DNA obtained from our theoretical model. A. number n of open bases as a function of time t for forces f ranging from 15.5 to 17 pN. B. magnification of the boxed region in Figure 3A after left hand rotation by 90 degrees. C. potential V(n), equal to the free energy the λ phage DNA with the first n base pairs open (V=0 corresponds to the fully zipped molecule) for $n < 450$ and force f=16.4 pN. The up and down arrows indicate, respectively, a local minimum in n=50 and two maxima in n=232 and n=327.

in the Bayesian theory framework [34]. The probability to have a sequence $\mathscr{S}$ given a time-trace $\mathscr{N}$ can be written as,

$$\mathscr{P}(\mathscr{S}|\mathscr{N}) = \frac{\mathscr{P}(\mathscr{N}|\mathscr{S})\,\mathscr{P}_0(\mathscr{S})}{\mathscr{P}(\mathscr{N})} . \qquad (4)$$

A prediction for a sequence is obtained by maximizing this probability over all possible sequences $\mathscr{S}$ (for a given $\mathscr{N}$). When no a priori information on the sequence is available, $\mathscr{P}_0(\mathscr{S})$ is uniform and the maximization of $\mathscr{P}(\mathscr{S}|\mathscr{N})$ over $\mathscr{S}$ reduces to that of $\mathscr{P}(\mathscr{N}|\mathscr{S})$. When, on the contrary, one does not aim to predict the content of a unknown molecule but to determine a mutation on a known molecule, $\mathscr{P}_0(\mathscr{S})$ contains the available a priori information on this sequence. $\mathscr{P}(\mathscr{N}|\mathscr{S})$ is defined from our dynamical model of the unzipping, and depends on the interval Δ between two measures.

We have first studied the unrealistic case of a very large bandwidth which allows us to follow all the base pair openings [33]. In this case the interval Δ between two measures is small with respect to the minimal base pair opening time (of about $1\mu s$ for an AT base pair). This value for Δ is not compatible with experimental limitations (e.g. due to the presence of the bead), indeed, nowadays, a realistic experimental value is $\Delta = 1ms$; however this limiting case can be studied in great details and some of the conclusions we obtained are generic in that they apply to the case of finite bandwidth. Once in silico unzipping data are generated with a Monte Carlo simulation a second program which ignores the phage sequence processes these data and makes a prediction for the sequence. This program is based on the Viterbi algorithm, widespread in information theory and in error correcting codes [34]. In Figure 8 we show the probability ε_n (full line, $R = 1$) that base number n (along the 5' to 3' strand) is wrongly predicted, plotted for the first 450 bases of the λ phage at a force of 16.4 pN. ε_n varies a lot from base to base with values ranging from 0 (perfect prediction) to 0.75 (random choice of a base value among A,T,C,G). Comparison with the potential $G(n, f = 16.4)$ shows that ε_n is small in the flat regions of the potential ($350 < n < 450$), or in the local minima, for example the base $n = 50$ that is preceded by 4 weak bases and followed by 4 strong bases (...TTTA-A-GGCG...). On the contrary, the bases that are not well predicted correspond to local maxima of the potential e.g. bases n=327 and 328 located between 7 strong bases and 7 weak bases (...GCCGCCG-TC-ATAAAAT...). For a force of 16.4 pN, 67% of bases are correctly predicted. The error increases with the force because, keeping the duration of the experiment fixed, the time spent by the fork on each base decreases. The fractions of mispredicted bases are 47% and 20% for forces equal to, respectively, 17 and 15.5 pN.

The quality of prediction can be enormously improved by collecting the data from repeated unzipping of the same molecule, and processing them altogether. Indeed by opening and closing the molecule several times the fork will surely go over the same portion of the sequence several times, and the signal over noise ratio increases. Figure 4A shows that the error sharply decreases when the number R of unzippings increases from 1 to 40. We have confirmed this numerical result with theoretical tools borrowed from the theory of disordered systems, of biased random walks and of large deviations in probability theory [34]. The error in predicting base n decreases very rapidly with the number R of unzippings,

$$\varepsilon_n \approx e^{-R/R_c(n,f)} \tag{5}$$

The typical number of unzipping to correctly infer the sequence $R_c(n, f)$ decrease with the force because the opening takes pleace slowly, with several passages in a same portion of the molecule before a barrier is crossed and a new region is visited. The fork then came back on the same, already open, base several times even in a single unzipping. $R_c(n, f)$ for the base n and at a force f can be calculated as the ratio of the decay constant at large force $R_c(i, f \geq 40)$ over the average number of openings of the base during a single unzipping, $\langle u_n \rangle$. $R_c(n, f \geq 40)$ depends on the sequence content on the bases $\{b_i\}$ around base n via the binding free energies $\{g_0(b_i, b_{i+1})\}$ (2) and can be exactly computed for any given sequence [33]. Stacking interactions generate first-neighboring coupling between bases and give rise to block of bases with strongly correlated value for $R_c(n, f \geq 40)$. The average number of openings of a base, $\langle u_n \rangle$, depends on the free energy landscape of the molecule (3), determined by the force and the sequence content, and can be computed for a given sequence [32, 33]. Our theoretical values for $R_c(n, f = 16.4 \text{ pN})$ ăare shown in figure 8B, and vary between 0.1 to 45 depending on the base pair index n. The agreement with the decay of ε_n from $R = 1$ to $R = 40$ is very good (Fig. 8 A).

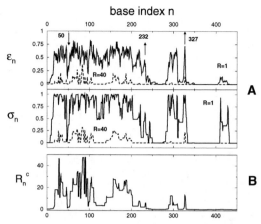

FIGURE 8. A. Probability ε_n that base n is not correctly predicted (top) and entropy σ_n (middle) for the first 450 bases of the λ phage DNA unzipped under a force of 16.4 pN. The full lines correspond to the predictions from data coming from one unzipping, the dotted lines from combined data from R=40 unzippings. B. Theoretical values for the decay constants $R_c(n)$ of the errors ε_n. As an example, the base 232 (arrow) is characterized by $R_c(232) = 6$, and is badly (respectively, well) predicted from the data from 1 (respectively, 40) unzippings.

Toward more realistic studies

We have made a step forward toward the analysis of real experimental data and have included in the inference analysis two major sources of instrumental limitations: the finite data acquisition bandwidth, and the elastic fluctuations of the unzipped DNA strands [32, 33]. In particular we have calculated the decay constant R_c of the prediction error with the number of collected data with a finite time interval Δ between two measures. We have also implemented an inference algorithm in the case of a small bandwidth or equivalently a large interval between two measures $\Delta \sim 10^{-3}s$. This algorithm, at difference from the Viterbi algorithm, is not guaranteed to find the maximum of (4), but correctly find back a sequence of 30 bases when 100,000 measures are collected (which corresponds with $\Delta \sim 10^{-3}$ s to some minutes of experiment). This result is in agreement with the theoretical value $R_c = 20000$ we found. Moreover in the Monte Carlo simulations the dynamics of the opening fork is directly monitored, while in real experiments the distance between the opened extremities is measured, and the ssDNA are not rigid linkers. We have extended our theoretical calculation by including the fluctuations of the ssDNA. As illustrated in Fig.9 due to the ssDNA fluctuations for each position of the opening fork n we have a distribution $A(x|n)$ of the ssDNA extremities distance x of the molecule. Moreover the variance of the distribution A increases with n because the longer the ssDNA is the floppier it becomes. Our finding is that even for a large bandwidth, $\Delta \simeq 10^{-6}s$, corresponding to the opening time of the base AT, the reconstruction procedure would imply a pre-treatment of the signal through a deconvolution of the measured n^a with the pseudo-inverse of A (see Fig. 9); moreover a number of unzippings that increases as $\sqrt{n}$ is necessary to reach the same inference performance

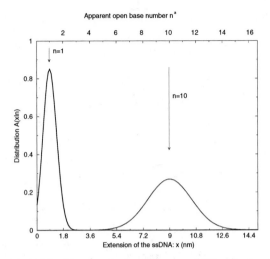

FIGURE 9. Distribution $A(x|n)$ of the extension x of the open ssDNA at fixed position n of the opening fork, for $n = 1$ and $n = 10$. The variance of the distribution (at a force of 16 pN) increase as $0.22\ nm^2 \times n$. The apparent value of the number of opened bases is n^a = closest integer to x/x_o, with $x_o = 0.9\ nm$, as shown on the top label of the x axis.

as in the absence of ssDNA fluctuations.

Conclusion

Micromanipulation experiments on single molecules of DNA are interesting approaches to extract information on DNA sequences. Among these experiments we have reviewed the unzipping of DNA under a mechanical action [4, 5, 7, 6, 8] or under translocation through a nano-pore [17, 18], and the observation of the sequence-dependent activity of an exo-nuclease [9, 10] and of a DNA polymerase [12, 13]. Even if these experiments cannot be nowaday competitive with the standard sequencing method, they could be useful for applications that require for example partial information on the sequence and where a complete sequencing is a waste of time and effort while a rapid single molecule screening would be more appropriate. Let us cite the analysis of the genetic variability of the genome, that is, the ability to locate in the genome of an individual the differences with respect to the average sequence (mutations, SNIPs, displacements of some portions of the sequence, ...) available in data bank e.g. the human genome bank. Two applications can be distinguished here. First the detection of known mutations. Secondly the search for genetic signatures corresponding to a given phenotype, signaling for instance a predisposition or resistance to some disease, or a particular reaction to some drug, ... This application should ultimately provide a better understanding of genetic and tumoral diseases, and lead to the development of personalized medicine.

So far these promising experiments have however not been combined with a concrete method for data analysis capable of extracting precise information on the sequence of

DNA molecules. The main difficulty comes from the intrinsically noisy nature of the measured signal. To make the single-molecule sequencing successful optimal experimental protocols have to be designed as well as inference methods capable of providing us with the most accurate information possible about the sequence.

REFERENCES

1. P.C. Turner, A.G. McLennan, A.D. Bates, M.R.H. White, *Molecular Biology*, Springer-Verlag (2000).
2. V. A. Bloomfield, D. M. Crothers, I. Tinoco J, Nucleic Acids Structures, Properties and Functions University Science Books, Sausalito, CA (2000).
3. S. Cocco, J.F. Marko, *Physics World* **16**, p. 37 (2003).
4. B. Essevaz-Roulet, U. Bockelmann, F. Heslot, *Proc. Natl. Acad. Sci. (USA)* **94**, p. 11935 (1997).
5. U. Bockelmann, P. Thomen, B. Essevaz-Roulet, V. Viasnoff, F. Heslot, *Biophys. J.* **82**, p. 1537 (2002).
6. J. Liphardt *et al. Science* **297**, p. 733 (2001).
7. S. Harlepp *et al. Eur. Phys. J. E* **12**, p. 605 (2003).
8. C. Danilowitcz *et al. Proc. Natl. Acad. Sci. (USA)* **100**, p. 1694 (2003).
9. A.M. Van Oijen, P.C. Blainey, D.J.Crampton, C.C. Richardson, T. Ellemberg, X. Sunney Xie, *Science* **301**, p. 123 (2003)
10. T.T. Perkins, R.V. Dalal, P.G. Mitsis, S.M. Block *Science* **301**, p. 1914 (2003).
11. C. Bustamante, Z. Bryant and S. B. Smith *Nature* **421**, p. 423 (2003).
12. GC Wuite, S.B. Smith, M. Young, D Keller, Bustamante C *Nature* **404**, p. 103 (2000).
13. B. Maier, D. Bensimon, V. Croquette *Proc. Natl. Acad. Sci. (USA)* **97**, p. 12002 (2000).
14. M.J. Levene, J Korlach J, SW Turner, M Foquet, HG Craighead, WW Webb *et al. Science* **299**, p. 682 (2003).
15. M. Rief,H Clausen-Schaumann, H Gaub *Nat Struct Biol* **6(4)** p. 346-9 (1999).
16. M. Singh-Zocchi,S. Dixit, V. Ivanov, G. Zocchi *Proc. Natl. Acad. Sci. (USA)* **100**, p. 7605 (2003).
17. A.F. Sauer-Budge, J.A. Nyamwanda, D.K. Lubensky, D. Branton *Phys. Rev. Lett.* **90**, p. 238101 (2003).
18. J. Mathé, H.Visram, V. Viasnoff, Y Rabin, A. Meller *Biophys. J.* **87**, p. 3205 (2004).
19. M. Zuker *Curr. Opin. Struct. Biol.* **10**, p. 303 (2000).
20. Santa Lucia Jr *Proc. Natl. Aca. Sci. (USA)* **95**, p. 1460 (1998).
21. S.B. Smith, Y. Cui, C. Bustamante, *Science* **271**, p. 795 (1996)
22. P. Cluzel , A. Lebrun, C. Heller, R. Lavery, JL Viovy, D Chatenay, F Caron *Science* **271** p. 792 (1996).
23. TR Strick, JF Allemand, D Bensimon, A Bensimon, V Croquette. *Science* **271**, p. 1835 (1996).
24. JF Marko *Phys. Rev. E* **57**, p. 2134 (1998).
25. S. Cocco, R. Monasson, J. Marko. *C.R. Physique* **3**, p. 569 (2002).
26. R.E. Thompson, E.D. Siggia. *Europhys. Lett.* **31**, p. 335 (1995).
27. S. Cocco, R. Monasson, J. Marko. *Eur. Phys. J. E* **10**, p. 153 (2003).
28. D.K. Lubensky, D.R. Nelson. *Phys. Rev. Lett.* **85**, 1572 (2000); *Phys. Rev. E* **65**, p. 031917 (2002).
29. U. Gerland, R. Bundschuh, T. Hwa. *Biophys. J.* **81**, p. 1324 (2001).
30. M. Manosas, F. Ritort, *Biophys. J.* **88**, p. 3224 (2004).
31. D. Marenduzzo *et al. Phys. Rev. Lett.* **88**, p. 028102 (2002).
32. V. Baldazzi *et al., Phys. Rev. Lett.* **96**, p. 128102 (2006)
33. V. Baldazzi *et al.,* Submitted to *Phys. Rev E* (2006).
34. D.J.C. McKay, Information Theory, Inference and Learning Algorithms, Cambridge University Press (2003).

Structural bioinformatics: advances and applications

Alejandro Giorgetti, Domenico Raimondo, Anna Tramontano

Istituto Pasteur – Fondazione Cenci Bolognetti and Department of Biochemical Sciences, University of Rome "La Sapienza", P.le A. Moro, 5 00185 Rome

Abstract. X-ray crystallography is the most used and effective technique for obtaining the structure of proteins and protein complexes. As of today, the x-ray structure of tens of thousand proteins is known and this number is continuously increasing, also thanks to the efforts of structural genomics projects aimed at providing representative examples of the protein structural space. In an x-ray diffraction experiment, crystals of the protein of interest are irradiated with x-ray, and interference effects give rise to a characteristic diffraction pattern. The amplitudes and phases of each reflection can be used to compute the electron density. However only, the intensity of the reflected waves can be measured, while their phase needs to be obtained by other means. One way to obtain the phase information is to use prior knowledge of the protein structure (search model). In some cases, the structure of a homologous protein or a model of the target protein can be sufficient to approximate the relative position of the atoms in the structure and allow the phases to be computed. This strategy is known under the name of molecular replacement. We have extensively investigated the relationship between the quality of a model and its usefulness as a search model in MR and discuss here our results.

Keywords: structural bioinformatics, molecular modeling, x-ray crystallography.
PACS: 61.82.Pv; 61.10.-i

DEFINITION OF THE PROBLEM

In a diffraction experiment, a crystal is irradiated with a particular X-ray wavelength and the resulting diffracted waves are collected on physical or electronic devices.. The electron density of the protein, i.e. the positions of the protein atoms, determines the diffraction pattern of the crystal, that is the magnitudes and phases of the X-ray diffraction waves, and vice versa, through a Fourier Transform function. In practice:

$$\rho(x, y, z) = \frac{1}{V} \sum_{hkl} \vec{F}_{hkl} = \frac{1}{V} \sum_{h} \sum_{k} \sum_{l} F(h, k, l) e^{-2\pi i (hx + ky + lz)} \tag{1}$$

where $\rho(x,y,z)$ is the electron density at position (x,y,z), $\vec{F}(h,k,l)$ is the vector describing the diffracted waves in terms of their amplitudes $F(h,k,l)$ and phases (the exponential complex term). The electron density at each point depends upon a sum of all of the amplitudes and phases of each reflection.

CP851, *From Physics to Biology; BIFI 2006 II International Congress*,
edited by J. Clemente-Gallardo, Y. Moreno, J. F. Sáenz Lorenzo, and A. Velázquez-Campoy

In x-ray crystallography, the resulting diffracted waves are collected on physical or electronic devices. However, in this passage from 3D to 2D all the information on the phases is lost and this is one of the fundamental problems of protein crystallography.

There are three approaches to solve the phase problem: direct methods, interference based methods and molecular replacement methods. The latter methods are based on the fact that the prior knowledge of a protein structure simplifies the solution of a different crystal form of the same molecule. In some cases, the structure of a homologous protein or a model of the target protein can be sufficient to approximate the relative position of the atoms in the structure and allow the structure factors to be computed[1]. Historically, it has been very difficult to decide a priori the quality of a model that is required for a successful molecular replacement experiment.

Here we describe a set of experiments that we performed to address the issue.

Protein structure modeling

Each possible conformation of an amino acid chain has a stability that depends upon the free energy change between its folded and unfolded states. Both entropy and enthalpy contribute to this free energy change. The first is associated with atomic interactions within the protein structure (dispersion forces, electrostatic interactions, van der Waals potentials and hydrogen bonding) while the entropy term describes hydrophobic interactions. Water tends to form ordered cages around non-polar molecules, such as the hydrophobic side chains of an unfolded protein. Upon folding of the polypeptide chain, these groups become buried within the protein structure and shielded from the solvent. The water molecules are more free to move and this leads to an increase in entropy that favours folding of the polypeptide.

Christian Anfinsen and his co-workers performed a series of experiments demonstrating that the native conformation of a protein is adopted spontaneously or, in other words, that the information contained in the protein sequence is sufficient to specify its structure [2]. These experiments immediately raised a fundamental problem known as the Levinthal paradox.

If the same native state is achieved by various folding processes, both *in vivo* and *in vitro*, we must conclude that the native state of a protein is thermodynamically the most stable state, i.e. the state where the interactions between the amino acids of the protein are the most energetically favourable with respect to all other possible arrangements that the chain can assume. But an amino acid chain has an enormous number of possible conformations (at least 2^{100} for a 100 amino acids chain, since at least two conformations are possible for each residue). It can be computed that the amino acid chain would need at least $\sim2^{100}$ picoseconds, or $\sim10^{10}$ years to sample all possible conformations and find the most stable structure. Levinthal concluded that a specific folding pathway must exist and that the native fold is simply the end of this pathway rather than the most stable chain fold [3]. There is a wealth of literature that addresses the Levinthal's paradox and, in general, the paradox can be solved by thinking of the folding process as a sequential process where the entropy decrease is immediately or nearly immediately compensated by an energy gain and that, in this hypothesis, the computed time scale of the folding process approximates what is observed in nature [4].

In order to compute the three-dimensional structure of a protein from first principles, not only we need to explore all possible protein's conformations. We also need to evaluate their free energy sufficiently accurately as to be able to identify the minimum energy configuration. This seems to be beyond our reach at present, as our calculations of the free energy are too approximate for the task to be accomplished [4].

Nevertheless, we do have methods for predicting protein structure, which are historically grouped according to the relationship between the target protein(s) and the proteins of known structure.

Comparative modeling can be applied when a clear evolutionary relationship between the target and a protein of known structure can be detected from sequence [5-7], while "fold recognition" methods can be applied when the structure of the target protein turns out to be related to that of a protein of known structure although the relationship is difficult, or impossible, to detect from the sequences [8-10]. When neither the sequence nor the structure of the target protein is similar to that of a known protein, we classify the methods as techniques for New Fold prediction.

Comparative modeling is based on the observation that evolutionary related proteins have similar conformations and therefore the experimental structure of a protein can be used as a starting model for that of other members of its evolutionary family [7]. However, as unrelated proteins often turn out to have similar structure, we can try and predict the structure of a protein even when we do not know the structure of an evolutionary related protein. We can reformulate the prediction problem and, rather than asking which is the structure of a target protein, we can ask (fold recognition methods) whether any of the known structures can represent a reasonable model for it, independently on the existence or detectability of an evolutionary relationship. several methods have been developed to evaluate the fitness function of a sequence and a structure. If the amino acid sequence of a protein does not reveal any evolutionary relationship with proteins of known structure and no fold recognition method proposes a putative structure with a sufficient level of confidence, we can make use of the observation proteins are composed of common structural motifs at the fragment or supersecondary structural level. This is exploited by "fragment based" methods that try and use fragments of proteins of known structure to reconstruct the complete structure of a target protein by using sequence-dependent local interactions as a bias to narrow the conformational space to be explored [11,12].

The CASP Experiment

The aim of the CASP (Critical Assessment of Techniques for Protein Structure Prediction) experiment is to evaluate the state of the art of prediction methods[13-17]. The experiment, which is repeated every two years since 1994, consists of several steps. In the first one structural biologists are asked to release the amino acid sequence of proteins, the CASP targets, whose structures are likely to be completed before the meeting. They are collected and made available to the "prediction community" whose members predict with whatever method they feel appropriate the structure of the target proteins and deposit their predictions before the experimental structures are released.

At the end of the prediction season (usually around the beginning of September), the targets, the models and a numerical evaluation of the quality of the predictions are made available to three assessors who are asked to critically evaluate the results, draw conclusions about the state of the art in the field of protein structure prediction and report on their analysis at a meeting, usually held in December. They also have the task to invite groups who obtained particularly good results or used interesting new methods to give a talk at the same meeting. Contributions by the assessors and selected predictors are published every time in a special issue of Proteins: Structure, Function, and Genetics.

CASP also evaluates the results provided by automated prediction servers, and compares them with those achieved by "human predictors". Of interest is the performance of the so-called meta-servers, i.e. of automatic servers that make use of several prediction servers, collect their results and try to pick the best answer by either selecting or combining the predictions.

Thanks to the continuous effort and drive of the prediction community, the role of CASP has gone beyond the evaluation of the state of the art in protein structure prediction: it has created a community that regularly meets and debates issues related to protein structure prediction developments and assessment, but also fosters new initiatives.

MOLECULAR REPLACEMENT WITH PROTEIN MODELS OF KNOWN QUALITY

A generally accepted "rule of thumb" is that molecular replacement is effective if a comparative model is used and if the latter has been built using as a "template" of protein of known structure sharing at least 40%-50% sequence identity with the unknown protein.

We decided to take advantage of the large set of models made available by the CASP experiment and ask the question of which are the requirements of models for being effectively used in MR. We used all available models and ran an automatic molecular replacement procedure, described in some detail later, using each of the models as a search model for molecular replacement.

Our results show that there is a correlation between the quality of the models and their suitability for molecular replacement, but that the traditional method of relying on sequence identity between the model and the template used to build it is not diagnostic for the success of the procedure [18].

The GDT-TS value [19], a distance based measure of correctness, is the measure that best correlates with the usefulness of the model in MR. In all cases, a GDT-TS above 84 is sufficient to guarantee the success of the procedure, regardless of the sequence identity between the target and template structure, of the method used for producing the model and of the structural class of the protein under examination. In our automatic procedure, models with GDT-TS below 80 are never successful. This implies that models built on the basis of a low sequence identity (above 25%) can be already sufficiently accurate for being used in the MR procedure, however minor differences in the quality of the models, not necessarily strictly correlated to the sequence identity between target and template, can make a substantial difference in the outcome (Table I). This underlines the fact that more efforts should be devoted to improve the initial model, since even minor improvements can be important for practical applications such as the one discussed here.

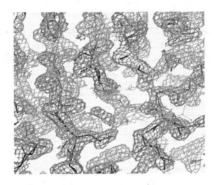

FIGURE 1. Example of the reconstruction of an electron density map using a CASP model as input for the automated molecular replacement procedure.

Models	Percentage of success
all models	71%
models with GDT-TS>84	100%
models with 80 < GDT-TS <84	33%
models with sequence id > 40%	46%

TABLE I – Statistics about the success of molecular replacement for CASP models.

AUTOMATIC MOLECULAR REPLACEMENT

In a real case, the results of the comparison of the model with the experimental structure are not available. It would be very useful to be able to identify beforehand features of the model that correlate with its ability to be used as a molecular replacement search model.

Furthermore, methods for building models of proteins are flourishing and trying each of the possible models of a protein in a molecular replacement experiment can be as time consuming as solving the phase problem with experimental techniques.

We have tested a fast approach to rapidly verify whether an automatically generated model can be successfully used for a molecular replacement experiment.

The strategy we employed is the following: we collected experimental structures as soon as they were made available from the Protein Data Bank [20] and sent them to a number of modelling servers, selected on the basis of their availability and fastness of response and clustered the resulting models. Our results indicate that a significant fraction of the structure could be reconstructed in a completely automatic fashion in more than 60% of the cases.

The reported percentage of success is a minimal estimate of the possibilities offered by modelling techniques in molecular replacement not only because of the aforementioned strict requirements that we imposed on the absence of manual intervention, but also because of some limitations in the programs used and, in some cases, of the available data. One aspect that requires further work is the treatment of multimeric proteins, which are rather difficult to handle in a completely automatic fashion.

FUTURE DEVELOPMENTS

The method we have developed might be useful to the crystallography community at large, at least as a first screening for potential molecular replacement techniques. We plan, therefore, to make it available as a server using a Web service based technology.

ACKNOWLEDGMENTS

This work was supported by Istituto Pasteur – Fondazione Cenci Bolognetti, and by the Bio-Sapiens Network of Excellence funded by the European Commission FP6 Programme, contract number LHSG-CT-203-503265.

REFERENCES

1. M. G. Rossmann. *The molecular replacement method.* **New York**: Gordon & Beach; 1972.
2. C. B. Anfinsen, W. F. Harrington, A. Hvidt, K. Lindstrom-Lang. *Biochim. Biophys. Acta* ;**17**:141-142 (1955).
3. C. Levinthal. *Journal de Chimie Physique et de Physico-Chimie Biologique* ;**65**:44-45 (1968).
4. A. Finkelstein, A. Gutin, A. Badretdinov. *Proteins* ;**23**:151-162 (1995).
5. A. Tramontano. *Methods (San Diego, Calif.)* ;**14**:293-300 (1998).
6. C. Levinthal. *Scientific American* ;**214**:42-52 (1966).
7. C. Chothia, A. Lesk. *EMBO J* ;**5**:823-826 (1986).
8. M. J. Sippl, S. Weitckus. *Proteins* ;**13**:258-71 (1992).
9. J. U. Bowie, R. Luthy, D. Eisenberg. *Science* ;**253**:164-70 (1991).
10. D. T. Jones. *Journal of Molecular Biology* ;**287**:797-815 (1999).
11. R. Bonneau, J. Tsai , I. Ruczinski, D. Chivian, C. Rohl, C. Strauss, D. Baker. *Proteins* ;**Suppl. 5**:119-126 (2001).
12. D. T. Jones, L. J. McGuffin. *Proteins* ;**53 Suppl 6**:480-5 (2003).
13. J. Moult, Z. A., K. Fidelis, T. Hubbard. *Proteins* ;**Suppl. 6**:334-339 (2003).
14. J. Moult, J. Pedersen, R. Judson, K. Fidelis. *Proteins* ;**23**:ii-v (1995).
15. J. Moult, T. Hubbard, S. Bryant, K. Fidelis, J. Pedersen. *Proteins* ;**Suppl 1**:2-6 (1997).
16. J. Moult, T. Hubbard, K. Fidelis, J. Pedersen. *Proteins* ;**Suppl 3**:2-6 (1999).
17. J. Moult, K. Fidelis, A. Zemla, T. Hubbard. *Proteins* ;**Suppl 5**:2-6 (2001).
18. A. Giorgetti, D. Raimondo, A. Miele, A. Tramontano. *Bioinformatics* ;**21**:72-76 (2005).
19. A. Zemla. *Nucleic Acids Research* ;**31**:3370-4 (2003).
20. F. C. Bernstein, T. F. Koetzle, G. J. Williams, J. Meyer E.F., M. D. Brice, J. R. Rodgers, O. Kennard, T. Shimanouchi, M. Tasumi. *Eur J Biochem* ;**80**:319-324 (1977).

Single-domain protein folding: a multi-faceted problem

Ivan Junier and Felix Ritort

Departament de Fisica Fonamental, Facultat de Fisica, Universitat de Barcelona, Diagonal 647, 08028 Barcelona, Spain

Abstract. We review theoretical approaches, experiments and numerical simulations that have been recently proposed to investigate the folding problem in single-domain proteins. From a theoretical point of view, we emphasize the energy landscape approach. As far as experiments are concerned, we focus on the recent development of single-molecule techniques. In particular, we compare the results obtained with two main techniques: single protein force measurements with optical tweezers and single-molecule fluorescence in studies on the same protein (RNase H). This allows us to point out some controversial issues such as the nature of the denatured and intermediate states and possible folding pathways. After reviewing the various numerical simulation techniques, we show that on-lattice protein-like models can help to understand many controversial issues.

Keywords: Protein folding, energy landscape, single-molecule experimental techniques, on-lattice heteropolymers
PACS: 87.14.Ee,82.20.Db,87.15.Cc

1. INTRODUCTION

Electrostatic forces, Van der Waals interactions, hydrogen bonds and entropic forces are the main elementary interactions that govern thermodynamics and kinetics of molecular interactions. In solution, electrostatic forces are mainly screened except at very short distances, typically on the order of the Angström. At these distances, chemical bonds of energies a few hundreds times larger than the thermal energy $k_B T$ tend to form at physiological temperatures $-k_B$ is the Boltzmann constant and T the temperature of the solvent. At physiological temperatures, the free energy of the hydrogen bonds involved in the formation of the protein secondary structures (α-helix, β-sheet) is around $2 \times k_B T$ [Fin02]. Also, the Van der Waals potentials that are responsible for the protein tertiary interactions are on the order of $k_B T$. Thus, in proteins (but also in nucleic acids), the thermal energy is comparable to the free energy of formation of non-covalent interactions. This leads to opposite effects. On one hand, it means that thermal agitation is the main source of intrinsic noise for biological processes [Mca99]. On the other hand, it suggests that thermal energy may be used as an energy source to trigger conformational changes and therefore induce mechanical work at the molecular level [Rit03]. However, in order to carry out specific tasks in a highly fluctuating environment, evolution, through natural selection, has favoured the formation of compact biological structures (DNA, RNA, protein) that are stabilized by multiple non-covalent bonds. RNAs and proteins are small enough to be activated by a small amount of energy available from ATP hydrolysis and, at the same time, stable enough to be biologically functional. DNA is a very long charged polymer but only a few number of base pairs

CP851, *From Physics to Biology; BIFI 2006 II International Congress,*
edited by J. Clemente-Gallardo, Y. Moreno, J. F. Sáenz Lorenzo, and A. Velázquez-Campoy
© 2006 American Institute of Physics 0-7354-0350-3/06/$23.00

are involved during transcription or replication processes. Furthermore, proteins, such as DNA polymerases or helicases, act *locally* on the DNA.

Proteins are ubiquitous molecules with a large variety of functions (regulatory, enzymatic, structural,...) [Pet04]. Regulatory proteins are involved in gene regulation processes, structural proteins (microtubules, actin filaments,....) give mechanical rigidity to the cell, transmembrane proteins regulate ion and water transport through membranes, etc... Proteins do not usually work alone. In some cases, a multiplex of several individual proteins participate in a common task, such as helicases and DNA polymerase proteins that coordinate their action during replication. In other cases, proteins are subunits of large molecular complexes such as the ribosome that consists of a patchwork of RNA and protein subunits.

During cell activity, proteins are continuously synthesized –and destroyed by protease proteins. Constitutive amino acids are transported by the transfer RNA and the ribosomes synthesize polypeptide sequences by matching the genetic code of the messenger RNA. Proteins have the remarkable ability to *fold upon a native structure*. This propensity was demonstrated in *in vitro* experiments by Anfinsen *et. al* [Anf73] in a denaturation/renaturation experiment of the Ribonuclease A protein in presence of urea. Subsequently, it has become clear that this is a general property of proteins since many experiments on different proteins have led to the same conclusion [Dob98]. The fast folding property is crucial since a protein becomes active only by adopting a specific thermodynamically stable structure (and in many cases by further forming specific complexes with small ligands). Furthermore, the diversity in protein functions is related to the diversity of protein structures. Folding of large proteins is helped by specialized biological machines, the so-called chaperons (GroEL-GroES, DnaK-DnaJ,...).

The fast folding property is not trivial. A random sequence of amino acids (and in extreme cases a single amino acid mutation of a good folder) leads to a polypeptide chain that behaves as a random coil without any specific structure [Dav94, Cre92]. To understand how proteins fold, different theoretical pictures have been proposed during the past twenty years [Fin02, Dil97, Onu04, Thi05]. Interestingly, recent single-molecule experiments [Bus03, Ben96, Bas97] that allow to investigate the biochemical processes at the molecular level [Bus03], in conjunction with increasingly powerful simulations, have refuted some of the theories and sharpened the big picture. Such a symbiosis between experiments, theory and numerical simulations have led to a better understanding about how biological machines work [Bus03]. For instance, one can now think of models that predict the native structure of a protein from its primary sequence [Bra03]. At a different level, biophysicists are able to observe *in real time* the action of single proteins acting on DNA, such as Gyrase, a protein that relaxes DNA supercoils [Gor06], and study it under different conditions (temperature, pH, tension, torsion,...).

This article is a short review about the folding/unfolding of small proteins. We first discuss the theoretical ideas that are nowadays used to tackle this problem. Next, we deal with the most recent experimental techniques that have provided important information about the folding mechanism of different proteins. We finish by reviewing the numerical techniques that are commonly used to investigate structural properties of the proteins during the folding transition. Our main goal is to present the basic notions necessary to understand the physics of the folding problem. The reader interested in a deeper understanding will find more detailed discussion in the proposed references of each section.

In section 2 , we discuss the energy landscape picture, a useful scheme that provides an intuitive idea of the folding propensity [Dil97, Onu97] but also that has led to quantitative tools useful to predict native and intermediate states [Onu04]. We further illustrate this approach by describing two simple models that show a protein-like behaviour. In section 3, we describe the main single-molecule techniques used to investigate individual proteins. We focus our discussion in underlining the differences between single-molecule force and single-molecule fluorescence experiments. To this end, we compare studies that have been carried out with the protein RNase H. In section 4, we review different numerical techniques such as molecular dynamics. We explore the use of coarse-grained models and give more details about generic lattice models that share protein-like properties. These models have the advantage of not being time-consuming, and allow to tackle general properties expected in single-domain protein folding. In particular, we address the questions of force-induced dynamics of a single-domain protein.

2. THE FREE ENERGY LANDSCAPE PICTURE

2.1. The Levinthal paradox

The structure of the native state of a protein is hierarchical. In the lowest level of description, a protein is described as a sequence of amino acids (residues) linked by peptide bonds. There are twenty different types of amino acids corresponding to different side chain groups –see Fig. 1. The residue sequence is called the primary structure. The formation of nearby hydrogen bonds between the amides and the carboxyl groups (Fig. 1) stabilizes the secondary structures mainly consisting of α-helices and β-sheets. The secondary structures are further stabilized by the tertiary interactions that are either hydrophobic interactions or disulfide bonds. Hydrophobicity results from the exposure of hydrophilic side chains to the solvent leading to the condensation of polar residues inside the core of the protein.

Each individual peptide group can have two conformations (Fig. 1). For an M-residue chain, one then roughly expects 2^M possible side chain configurations. Assuming that the minimal timescale for a stereoisomeric conformational change is about one picosecond, then the total time required for visiting all the configurations should be $\sim 2^M$ps $\sim$ 10^{10}years for a 100-residues protein [Fin02]. This crude approximation shows that the folding process can not consist of a random search in the protein configurational space [Lev68]. On the contrary, the *energy landscape*, i.e. the energy surface as a function of the configurational parameters (the degrees of freedom) –see Fig. 2– , is biased toward the native structure as depicted in Fig. 2. Within the ideal picture of Fig. 2, at sufficiently "low temperatures" (when k_BT is on the order of the formation energy of a native contact [Zwa92]) the *free energy landscape* is biased by the energy gradient leading to downhill motion and collapse towards the native structure [Dil97, Onu04, Onu97]. This situation corresponds to a perfect funnelled landscape [Onu04]. The underlying mechanism that leads to such a smooth landscape is referred as "minimal frustration" [Onu97]. A minimally frustrated structure is a structure for which the intramolecular interactions are not in conflict with each other leading to a smooth landscape

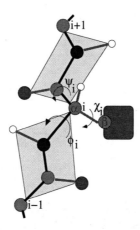

FIGURE 1. A protein is a chain of amino acids linked by peptide bonds (the peptide units are outlined by he parallelograms). The side-chain (outlined in green) defines the residue (amino acid + side-chain). Twenty types of side-chain exist, the most simple being an atom of hydrogen that is called glycine. In this case, there is no β-carbon. The peptide units are planar, due to the sp^2 hybridization type of the N-C bond. Different χ angles correspond to different conformations of the side-chain. Two conformations (*cis* and *trans*) are possible for the peptide unit, depending on the positions of the oxygen and the hydrogen at the tops of the parallelogram. Carbon, oxygen, nitrogen and hydrogen atoms are respectively represented in grey, red, blue and white. The backbone structure is highlighted in black.

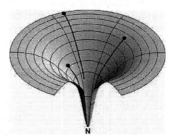

FIGURE 2. Artistic cartoon of a perfect funnelled landscape. The vertical axis counts for the energy and the horizontal plane for the degrees of freedom of the polypeptide chain. Taken from [Dil97].

as in Fig 2. The concept of a funnel is not only qualitative but also quantitative. The simplest way to design a perfect funnel is by considering interactions that only stabilize the native structure [Onu04]. By following this strategy and using a coarse-grained description of proteins (e.g. the Gō model (see section 4.1)), excellent predictions of native structures and even intermediate states have been observed [Onu04]. However, the perfect funnelled landscape is not so general. A rough energy landscape with many local minima and saddles corresponding to configurations with various degrees of stability, is more appropriate. Within this picture, misfolded behaviour results from the competition between local minima that are close to the native state [Onu97]. For a detailed discussion about the energy landscape, see the review by Onuchic *et. al* [Onu97].

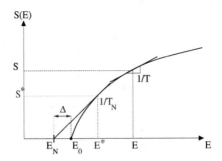

FIGURE 3. Dependence of the entropy in a random energy protein. In the pure random energy model (without the native state at energy E_N), the mean value of the energy E and the entropy S are given by the coordinates at which the derivative of $S(E)$ is equal to $1/T$. In the composite model (i.e. with the native state), the free energy equality between the native state and the denatured states implies $1/T_N = S^*/(E^* - E_N)$. T_N is the transition temperature where the native state and the denatured state are equal likely. Δ is the energy gap.

2.1.1. Mixing stochasticity and determinism

The energy landscape picture, by definition, leads to non-specific folding pathways from the many denatured (i.e. non native) conformations to the folded native state. This scheme has been opposed for a long time to the very first scenarios aiming at explaining the folding property. According to the latter, the folding process is a specific mechanism whose dynamics is sequential, which leads to a unique folding pathway [Fin02], by opposition to the stochastic nature of the energy landscape approach [Dil97, Onu97, Fin02]. These two view points are not contradictory but rather describe mechanisms at different levels. For instance, Lazaridis and Karplus [Laz97] have studied 24 unfolding trajectories of a small protein (chymotrypsin inhibitor 2), using molecular dynamics. They have observed large statistical fluctuations in the gyration radius of the successive structures during the unfolding process, in agreement with the energy landscape picture. On the other hand, some specific events, such as the destruction of tertiary contacts, were found to be specifically ordered in time [Laz97].

2.2. Thermodynamics

In most cases, small globular proteins fold following an all-or-none process, just as do small RNA hairpins. The origin of this cooperative effect lies in the fact that the native state has a very low entropy. Thus, the transition from the denatured state (with high entropy) to the native state is generally accompanied by an entropy jump or, equivalently, a peak in the specific heat as observed in bulk denaturation experiments [Fin02]. In the following, we discuss two simple models describing the transition between a high entropy phase and a very low entropy native phase.

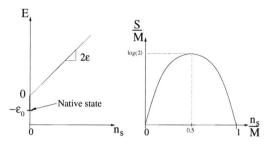

FIGURE 4. The Zwanzig picture. n_s is the number of non-native contacts. Left: the potential $E(n_s)$ is given by $E(n_s) = 2\varepsilon n_s$ when there are non-native contacts ($n_s > 0$). At the native state, $E(0) = -\varepsilon_0$. Therefore, the energy gap between the native state and the lowest denatured states is equal to ε_0. Right: the entropy as a function of n_s/M for large M. In this limit, $S(x = n_s/M)/M = (x-1)\log(1-x) - x\log x$.

2.2.1. The entropy crisis avoided.

In the glass phenomenology, it has been hypothesized that there is a finite temperature at which the configurational entropy ([total entropy] − [vibrational entropy]) vanishes [Edi96]. This has been called "the entropy crisis" by Kauzmann [Kau48]. The simplest model describing the entropy crisis is the random energy model (REM) [Der80]. In this model, the entropy is a quadratic function of the energy that vanishes at an energy E_0, i.e. there is no state with an energy below E_0. At equilibrium, the free energy corresponds to the point in the entropy curve, $S(E)$, at which its tangent is equal to $1/T$ –see Fig. 3. As a consequence, as $T \to T_0$ where T_0 corresponds to the energy E_0, the entropy *continuously* vanishes. The point at $E = E_0$ defines the glass transition. By incorporating into this model a native state with an energy $E_N = E_0 - \Delta$ (Δ is the so-called energy gap), one gets a first-order transition [Onu97], between a high entropy state and the native state, at a temperature $T_N = (E^* - E_N)/S^*$. At T_N, the free energy of the denatured state, $F^* = E^* - T_N S^*$, is equal to the native one, $F_N = E_N$. It can be shown that the glass transition is avoided if the energy gap is much larger than E_0/M [Bry89, Onu04, Fin02], M being the number of residues. The transition then becomes first order.

2.2.2. Funnel-driven transition

The energy funnel is akin to the minimal frustration property. The simplest model exhibiting a perfect funnelled landscape is the Zwanzig model [Zwa95]. It can be thought of as a spin model where the energy of a given spin configuration $\mathcal{C} = \{s_1...s_M\}$ (we consider a set of M spins) reads $E = \varepsilon \sum_i |s_i - s_i^N|/2 - \varepsilon_0 \delta(\mathcal{C} - \mathcal{C}^N)$ where $\mathcal{C}^N = \{s_1^N...s_M^N\}$ is the native configuration. The parameters ε and ε_0 are positive energies related to the gradient of the funnel and the native gap Δ respectively –see Fig. 4. The energy of this system can be explicitly written as a function of the number n_s of spins that differ from the spins in the native configuration [Zwa95]. We will call $M - n_s$ the number of native contacts. Let us now consider a single-spin dynamics with Metropolis rules. In this case, at any time there are only two kinds of elementary moves: a spin-flip can lead

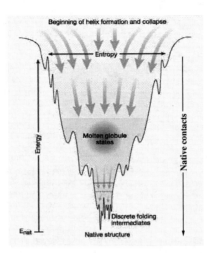

FIGURE 5. The energy landscape picture. In general, the energy landscape of a protein is rugged and funnelled, i.e. with an overall gradient that is oriented toward the unique native state. The landscape is stratified according to the energy of the configuration, or according to the percentage of native contacts. Misfolded states are the local minima closed to the native state. Adapted from [Onu97].

to a new native contact ($n_s \rightarrow n_s - 1$) or to a new non-native contact ($n_s \rightarrow n_s + 1$). Since there is no interaction between the spins, there is no conflict between the interactions, which means that the probability to have a native contact at a site i does not depend on the configuration of the other spins. A set of non-interacting constituents that feel a time-independent local potential is therefore the simplest example of a minimally frustrated system since there is no frustration at all.

One can write a master equation for the probability density of the number of contacts n_s at a given time, where n_s represents a reaction coordinate [Zwa95]. The thermodynamic potential $E(n_s)$, that reflects the funnel-shape, is linear in n_s and has a gap at $n_s=0$ (Fig. 4). As occurs in any folding transition, there is a competition between the entropy, that favours denatured states (non-native contacts), and the potential energy $E(n_s)$ that biases the system towards the native structure. In the limit of large number of spins, one finds a temperature transition, $T_N = \varepsilon$, at which the probability of being in the folded/unfolded state is $1/2$ [Zwa95].

2.2.3. The potential of mean force

Taking into account the ruggedness of the energy surface, the funnelled shape of the *potential energy landscape* is usually represented as in Fig. 5 [Onu97]. However, the *free energy landscape* is more suitable to discuss a possible thermodynamic transition. Contrary to bulk experiments, in which measurements lead to ensemble averaged quantities, single-molecule experiments allow to compute the free energy as a function of reaction coordinates such as the molecular extension, a quantity that is related to the number of

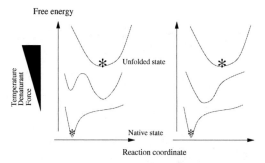

FIGURE 6. The free energy landscape, or equivalently the potential of mean force. Two scenarios are usually observed. Left: the transition is first order with a coexistence phase between the native state and a denatured state. Right: The transition is continuous. The free energy projection always shows a single well that drifts toward the native state as the folding conditions become more appropriate (bottom panels). The * indicate the denatured states and the native states.

native contacts. This is usually called the potential of mean force. In the following, we discuss situations where the free energy is projected along a single reaction coordinate. Nevertheless, it must be stressed that many aspects of the folding kinetics can not be understood without considering more than one reaction coordinate (see e.g. [Lee00]).

The free energy landscape approach calls for two general situations (Fig. 6) that have been experimentally observed. On one hand, some single-domain protein experiments have revealed a continuous phase transition between denatured states and the native state [Mun04]. In this case, at any temperature, the free energy landscape is composed of a single well. The minimum of the well drifts towards the native structure as one lowers the temperature or decreases the denaturant concentration. On the other hand, most of the single-molecule (and bulk) experiments involving single-domain proteins have revealed a first order transition [Fin02, Bak00]: the free energy profile consists of two wells with minima corresponding to the denatured and the native states –see Fig. 6. As in any first-order transition, the system goes from a denatured state to the native state by passing through a coexistence phase. In bulk experiments, this leads to the presence of proteins that are denatured and proteins that are in a native state. From the single-protein point of view, this suggests cooperative switches between the native and the denatured states as reported in Fig. 7.

2.3. Kinetics

The energy landscape represents a useful picture to understand the existence of mis-folded structures and, more generally, the folding kinetics of a single-molecule. By considering the overdamped motion of a particle along a potential of mean force (free energy projection), one implicitly makes the assumption that all the degrees of freedom orthogonal to the reaction coordinate locally equilibrate [Ris89]. This may not be true in specific non-equilibrium conditions, such as low temperatures where proteins show glassy behaviour [Onu97].

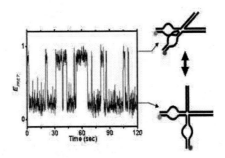

FIGURE 7. Example of cooperative transitions observed in a single-molecule experiment [Tan03]. Left: the vertical axis represents the FRET efficiency (see section 3.1) that reflects the state of the biomolecule (here an RNA molecule hairpin ribozyme shcematically represented on the right part of the figure). One can see that the molecule switches between the native state (upper configuration on the right) and the denatured state (lower configuration). Taken from [Ha].

2.3.1. *Two-states and downhill kinetic scenarios*

Kramers theory allows to derive dynamical properties related to the diffusion motion of a particle along a one-dimensional landscape [Zwa01]. In particular, the mean first-passage time between the denatured and the native states can be computed to extract the effective free energy barriers. A two-states description of the Kramers' problem models the dynamics in terms of activated events across a free energy barrier and represents the simplest description of a cooperative all-or-none transition. This approach is potentially useful to understand single-molecule force experiments, e.g. in the force unfolding of single RNA molecules [Rit02]. When the position of the transition state moves along the reaction coordinate by changing the external conditions (temperature, denaturant concentration, stretching force), known as the Hammond behaviour [Ham55], an extended two-states description with a mobile barrier can be applied [Man06].

The existence of free energy barriers that make the transition all-or-none is usually attributed to the asynchronous compensation between energy gain and entropy loss [Onu97]. However, continuous transitions have been also observed in recent experiments [Gar02]. These transitions can then be thought as a limiting case of the two-states model where the free energy barrier becomes comparable to $k_B T$. It is then more convenient to see the folding as a downhill process [Mun04, Gar02]. Notice that the ideal funnel picture of Fig. 2 actually suggests a compensation between entropy (given by the radius of the funnel) and energy (given by the depth of the funnel).

The two-states transition between either the native state and a random coil (with no native contacts) or the native state and a molten-globule structure (with numerous native contacts) has been for long time a well accepted scenario for single-domain proteins [Fin02]. However, this has been disputed in recent numerical studies on the lyzozyme (1HEL) [Fit04]. This is a single-domain protein known to exhibit at room temperature a first order transition between the native state and a pure random coil as the concentration of denaturant (guanidine dihydrochloryde) is increased [Tan66]. In fact, it has been shown that even in the presence of many native contacts (more than

FIGURE 8. The RNase H protein structure. The colourful stars represent the dyes that are chemically attached to the protein. These dyes are used in the florescence techniques, namely the FRET measurement –see text. r, the distance between the dyes, is directly related to the molecular extension of the protein. Adapted from [Cec05].

90%) the gyration radius and the end-to-end distance are well described by the Gaussian random coil model [Fit04]. Therefore, standard bulk experiments may not provide enough information to distinguish a random coil state from a native-like state. In the next section, we describe single-molecule techniques that might resolve this controversy by addressing new interesting questions.

3. SINGLE-MOLECULE EXPERIMENTS

In this section, we review the principal techniques used to investigate proteins in single-molecule experiments. We focus our discussion on two of them: fluorescence spectroscopy and force measurements. We compare the results obtained with these techniques in the RNase H protein and discuss whether force may induce folding pathways different than those of thermal folding.

3.1. Fluorescence techniques

Three-dimensional native structures can be determined in solution by nuclear magnetic resonance (NMR) spectroscopy or in crystal forming proteins by X-ray crystallography. For small globular proteins, the two measurements give generally the same result, showing that the native structure is a highly compact structure in solution. Such techniques are inappropriate for studying the structure of the transition state and the denatured states. Indeed, it is impossible to crystallize fluctuating states and NMR measurements average out conformational fluctuations. Nevertheless, some bulk techniques, e.g. small-angle X-ray and neutron scattering, have provided precious information about quantities such as the radius of gyration [Mil02]. In particular, it has been shown that random coils are not the most general denatured state [Mil02], even at high denaturant concentrations. Recently, these results have been unambiguously confirmed by using single-molecule fluorescence techniques.

Fluorescence techniques are based on the so-called Förster resonant energy transfer (FRET). A green fluorescent donor dye and a red fluorescent acceptor are chemically attached to the end residues of the protein (Fig. 8). The donor is excited by a well-tuned laser and further relaxes by emitting a fluorescent light that can be detected by a spectrophotometer. The acceptor is chosen such that its absorption spectrum overlaps the emission spectrum of the donor. As a consequence, a non-radiative energy transfer between the chromophores may decrease the intensity of the donor by enhancing the emission of the acceptor. The (FRET) efficiency of energy transfer between acceptor and donor depends on their distance r, and hence on the protein extension, through the simple relation

$$E = \frac{1}{1 + (r/R_0)^6} \tag{1}$$

where R_0 is a characteristic parameter of the pair of dyes. On the other hand, the efficiency E can be directly related to the emitted intensities by the dyes:

$$E = \frac{I_A}{I_D + I_A} \tag{2}$$

where I_D and I_A are the intensities emitted by the donor and the acceptor.

As a result, a quantitative spectral detection of the dyes gives information about the conformation of the protein. One may even think of placing the dyes at different locations in the protein to get further structural information. In this spirit, this technique has been recently used to investigate specific conformational changes during biological processes. For instance, it has been used to follow the different steps of protein synthesis in the ribosome [Bla04]. This is very important to understand the mechanism responsible for the so exclusive codon/anticodon recognition by the transfer RNAs.

3.1.1. The RNase H protein

Let us now focus on folding studies in a small single-domain protein, the 155-residue RNase H protein. The native structure of this protein is well known and is shown in Fig. 8. Under appropriate folding conditions, several (bulk) studies have shown that the folding is preceded by a fast collapse to a compact structure presumably stabilized by a central nucleus [Ras99, Bal99].

Nienhaus and co-workers have used the above fluorescence technique to investigate several structural properties of the RNase H protein [Kuz05, Kuz06]. To determine the spectral properties of the dyes, they fix an ensemble of proteins on a glass surface. A FRET histogram is obtained by "counting" the number of proteins with efficiency E (see Eq. 1). By varying the concentration of denaturant, e.g. the guanidine dihydrochloride, they have monitored the cooperative transition between the native and the denatured states. From these measurements, it is then possible to extract the corresponding potential of mean force (free energy landscape) along a reaction coordinate that is related to the compactness of the protein –see Fig. 9. This coordinate actually characterizes the propensity of the molecule to let the solvent enter. Notice that these curves could be the curves of any single-domain protein. Interestingly, the folding free energy changes

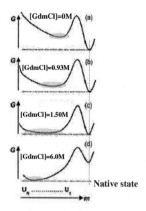

FIGURE 9. Free energy profiles of the RNase H as a function of the denaturant concentration [*GdmCl*]. Upper panel: the thermodynamically stable state is the native state. Bottom panel: the thermodynamically stable state is a denatured state. The abscissa is the so-called cooperativity parameter and is related to the propensity of the state to let the solvent enter into the molecule. $\{U_i\}_{i=1N}$ is a set of denatured states structurally close to the native state, the closest being U_1. Taken from [Kuz06].

as one varies the concentration of denaturant. This raises several questions: to what extent the denatured state at low denaturant concentration is different from the denatured state at high concentration? Is the transition between the high-denaturant state and the low-denaturant state of the same type as the continuous transition discussed above? Does the high concentration denaturant state have a residual structure reminiscent of the native state? Fluorescent studies by the Nienhaus group have shown that even at high denaturant concentration, the denatured state was composed of non-random structures [Kuz05, Kuz06]. This study, in conjunction with the numerical simulations of the folding of lyzozyme (1HEL) [Fit04], cast serious doubts about the true nature of the denatured state, an issue that has been an experimental challenge for a long time. Moreover, it is also a difficult problem from the point of view of numerical simulations because of the huge number of accessible configurations.

The RNase H results [Kuz05] suggest the existence at low denaturant concentration of a well-defined compact structure different from the native state. As discussed above, this structure was expected from earlier stop-flow kinetic experiments in which RNase H often showed the accumulation of compact structures during the dead-time of the measurement. Interestingly, force experiments applied to the same molecule have also shown the existence of a well-defined intermediate state that coexists with both the denatured and the native states [Cec05].

3.2. Force measurements

Force measurements on a single molecule have been first realized on a double-stranded DNA [Bus03]. A fluid flow and a magnet were used to stretch the molecule that was attached to micron-size beads. The measurement of the molecular extension

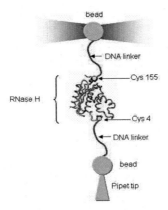

FIGURE 10. Setup of the force measurement in the single-protein RNase H experiment. DNA linkers between the beads and the protein are inserted in order to be able to manipulate the protein. The DNA linkers are chemically attached to the end of the proteins via the insertion of a cystein side-chain. The bead at the top of the pipette is held fixed by air suction and the other bead is trapped in the optical well.

of DNA has then revealed unexpected mechanical properties, such as the overstretching transition [Bus03]. Subsequently, different studies have been realized [Bus03, Ben96] in order to investigate the behaviour of DNA under torsional strain (using magnetic beads that can be rotated by magnets) [Str00, Str96], the DNA and RNA unzipping process (using optical tweezers) [Boc02, Coc03], the DNA packaging problem [Smi01], DNA/protein interactions [All03] or DNA condensation [Rit06]. As a consequence, single-molecule force experiments have contributed a lot toward our understanding of the cell machinery. Single-molecule force measurements (on RNA) [Col05] have been also used to test non-equilibrium theories in statistical physics and to recover folding free energies in RNA molecules. In this spirit, biomolecules appear to be ideal systems to explore the thermodynamic behaviour of small systems and to test non-equilibrium theories in statistical physics [Bus05].

Recent nano-manipulation of single protein molecules using the atomic force microscope (AFM) [Fis99] have provided direct evidence for sequential unfolding of individual domains upon stretching [Fis00, Kel97, Rie97, Sch05]. However, optical tweezers are more appropriate to study the unfolding/folding dynamics of small single-proteins (and small RNAs). In fact, the folding free energies of such biomolecules are on the order of $100k_BT$ at room temperature. Considering a typical gain in extension of $\Delta x \sim 10 - 20nm$, between the native state and a stretched state, a mechanical energy $f\Delta x \sim 100k_BT$ is provided by a stretching force on the order of $10pN$, which is in the ideal working range of optical tweezers [Lan03]. In contrast, the AFM technique is useful to investigate forces above a few tens of pN [Fis99] but can not reach forces $\sim pN$ mainly because of the high spring constant of the cantilever.

Typical optical tweezers experiments use micron-sized glass chambers filled with water and two beads. The protein is chemically labelled at its end and polystyrene beads are chemically coated to stick to the ends of the labelled molecule. Because proteins are too small to be manipulated with micro-sized beads, a tether consisting of a double

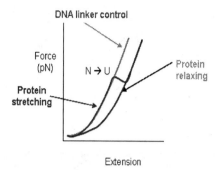

FIGURE 11. Typical force-extension curve (FEC) during the unfolding/refolding ramp force protocol of a single-domain protein. The undolding of the protein corresponds to the extension jump (in red). The rest of the curve is well described by a worm-like chain model that models the extension of the linkers when the protein is still folded, and the extension of the (linkers + unfolded protein) when the protein is unfolded.

stranded DNA is inserted between the beads and the molecule that acts as a polymer spacer –see Fig. 10. This prevents Van der Waals forces between the beads and the protein and allows a direct manipulation of the protein. One bead is then held fixed by air suction on the tip of a glass micro-pipette, the other is trapped in the focus of a laser beam. When the bead deviates from the focus a restoring force acts upon the bead, the principle being the same by which a dielectric substance inside a capacitor is drawn inwards by the action of the electric field. To a good approximation, the trap potential is harmonic. Thus, knowing the trap stiffness, it is possible to apply mechanical force (by moving the bead) and to observe in real-time the force-extension curves (FEC). In the FECs, the force acting on the molecule is represented as a function of the end-to-end distance between the two beads. The cooperative opening of the proteins is characterized by a jump in the extension of the molecule –see Fig. 11. By studying the stochastic properties of the FECs, one is able to recover the distance from the native state to the transition state and map the free energy landscape as a function of the molecular extension. The folding and the unfolding rates can also be determined [Man06, Sch05].

3.2.1. The RNase H protein

In the case of the RNase H protein, Cecconi *et al* [Cec05] have shown that mechanical forces can stabilize an intermediate state. We use the word "stabilize" since the intermediate state corresponds to a local minimum of the free energy landscape projected along the end-to-end distance (Fig. 13) that is well separated from the unfolded state and the native state by all-or-none transitions (Fig. 12). At constant force, three regimes can be distinguished depending on the value of the force: at high force, the molecule is fully stretched and no native residual contacts are present; at low force, the protein is in its

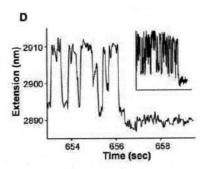

FIGURE 12. Extension trace of the RNase H protein at constant force ($\approx 5.5\,pN$). In the first part, we see successive all-or-none transitions between the intermediate state and the unfolded state (fully stretched). Then, a transition occurs between the intermediate state and the native state showing that the intermediate state is on-pathway. Figure taken from [Cec05].

native compact state; in-between, there is an intermediate state with a partial number of native contacts that are formed. Three states, instead of two, coexist: the stretched, the native and the intermediate compact states (see Fig. 13). A statistical analysis of the breakage force and measurement of the rip extensions have led to an extrapolated zero-force intermediate free energy that correlates well with that of the early compact structure that forms in bulk experiment [Ras99].

3.3. Comparing force and FRET measurements

A comparison between the folding/unfolding study of RNase H *with* and *without* force raises interesting questions relative to the structure of proteins: 1) Under which conditions do we expect that the early molten-globule state that forms at zero force is the intermediate state stabilized by mechanical force? 2) FRET measurements have revealed a hierarchical structure of RNase H in the denatured state [Kuz06]. Is the stabilization of the intermediate state related to this observation? 3) More generally, is the stabilization of an intermediate state a signature of a specific folding mechanism? Such questions can be actually addressed in numerical simulations of simpler models [Jun06].

The force measurements in RNase H also raise questions about the on/off pathway nature of the intermediate states, an issue that we discuss in the next paragraph.

3.4. Probing the nature of the intermediate states

Let us consider a system with a free energy landscape showing three well-separated minima –see Fig. 13. One might wonder whether a diffusive dynamics along this profile fairly reproduces the observed dynamical behaviour in the single-molecule experiment (Fig. 12). In such case, by starting from any state in the intermediate region and preventing the system from going to the stretched region, the molecule should be able to fold to

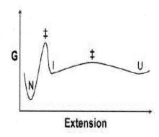

Extension

FIGURE 13. Free energy (G) in RNase H protein projected along its molecular extension. Three regions, that are delimited by the signs ‡, can be defined: the native region where the minimum corresponds to the native state (N), the intermediate region where the minimum corresponds to the intermediate state (I) and the stretched region where the minimum corresponds to the unfolded state (U).

the native state. Misfolded (off-pathway) states are those that can not lead to the native state without unfolding back to the stretched states.

In the RHase H force study, the extension trace of Fig. 12 suggests that folding indeed takes place *via* the intermediate states [Cec05]. However, we can not discard additional states lying at the same coordinate than the intermediate state and that can not lead to the native state without unfolding back to the stretched state. We can propose an experimental protocol to quantify the fraction of off-pathway states with respect to on-pathway states. Each time the system jumps to the intermediate state, we suddenly relax the force to a lower value. On-pathway states should quickly lead to the native state without unfolding back to the stretched state. For off-pathway states however, it is expected a first unfolding event to the stretched state and, most likely, an extremely large folding time as compared to the typical on-pathway folding time. We give a numerical example of such a protocol in the next section.

4. NUMERICAL SIMULATIONS

From an experimental point of view, it is still a challenge to get atomic structural information of intermediate states, transition states (corresponding to the maxima of the projected free energy landscape), or denatured states. One could think of a fluorescence technique using dyes attached to different residues of the proteins. However, the presence of the chromophores inside the molecule is likely to impede the correct folding of the molecule or to modify the real structure of the expanded states. So far, the best way to characterize non-native states has been to resort to numerical simulations. The latter can be divided into three classes:

1. *Molecular dynamics.* It takes into account all (or almost) the atomic details of the molecule and the solvent can be explicitly or implicitly treated [Kar02]. The folding pathways are determined by simulating trajectories in quasi reversible conditions. The technique is time limited because only nanosecond long unfolding trajectories can be obtained whereas folding of real proteins occurs mostly in microsecond timescales. High temperatures or mechanical forces are then usually used to

accelerate the unfolding trajectories [Kar02]. The main argument is that different conditions induce different timescales but not different mechanisms. Current improvements in this field have been achieved thanks to the development of more and more accurate interatomic potentials in different environments [Cor95].

Interestingly, the original technique has been also adapted to simulate a large set of short trajectories (starting from random configurations) of a designed small protein (23 residues) at room temperature [Sno02]. A non-negligible amount of very fast folding trajectories ($\sim 20ns$) has been observed whereas in experiments the mean folding time is on the order of microseconds. This shows, as expected for two-state cooperative proteins, that the folding step is very short but the whole folding mechanism is slowed down due to the presence of many possible denatured configurations. As mentioned in section 2.3.1, this comes from an asynchronous compensation between entropy and energy.

2. *Coarse-grained models.* Within this scheme, one reduces the all-atom description to a mesoscopic description by neglecting details of the polypeptide chain [Bak00, Zho99, Vei97]. The parameters of the mesoscopic description are obtained by a close comparison with experiments. Different levels of simplification are usually taken into account. Perfect funnel models including only interactions that stabilize the native structure have led to excellent predictions of native and transition states [Bak00, Onu04]. Less restrictive Gō models lead to a more refined statistical description of folding trajectories [Cle03, Onu04]. At the end of this section, we describe details of such simulations and discuss the issue in the presence of mechanical forces. Notice that modelling of the solvent is also essential to understand protein folding. Within this scope, Gō-like potential including a (de)solvation potential can be taken into account to model the expulsion of water molecules from the protein core [Onu04].

3. *Protein-like models.* The purpose of these models is not to study the folding mechanism of some specific proteins but rather to give general insights about the folding dynamics. They are used to determine whether the dynamics is related to the (hetero-)polymeric properties of proteins, such as the native state geometry or the contour length of the chain. They are also useful to investigate the folding behaviour in presence of specific external conditions (temperature, denaturant concentration, stretching force...). Although there may have qualitative differences between on-lattice and off-lattice models (see e.g. [Pan98]), most of the studies have been done with models defined on a lattice, the main reason being the possibility to simulate large molecules during long times. In the following, we review some of these models and show how to incorporate mechanical forces. In particular, we show that these models can be useful to tackle the problem about the on/off pathway states.

4.1. Protein-like behaviour of simple models

4.1.1. Hydrophobic-Polar models

Heteropolymers on a lattice with simple hydrophobic-polar interactions between non-adjacent monomers are the simplest models that show a protein-like behaviour. In the *HP* model [Cha89, Cha94], a diblock copolymer chain composed of hydrophobic (*H*) and hydrophilic, equivalently polar, (*P*) monomers is considered on a square lattice. Only the interactions *HH* are energetically favourable, the so-called Gō interaction. Specific sequences (HPPH...) then lead to protein-like behaviour and have been used to exhaustively explore the underlying energy landscape [Cha94]. Interestingly, the folding mechanism has been shown to be reminiscent of small single-proteins. Indeed, under appropriate folding conditions, the extended chain quickly condensate into a rich *HH*-bonds structure.

Since non-native *HH* bonds are present, the molecule further needs to break *HH* bonds to get closer to the native state. This stage is similar to the exploration of a non-native compact structure set that precedes the fast downhill step. Such a process actually goes accompanied with an expansion of the structure in order to allow local conformational changes of the polymer, a behaviour that has been experimentally observed [Cha94]. More quantitatively, a recent study of this model [Kac06] has pointed out that the folding rates may not be correlated to the thermodynamic properties of the molecule, such as the value of the energy gap and the structure of the native state. It rather suggests that the folding rates are well correlated with the number of local energy minima, i.e. the former decreases as the latter increases. These results are in good agreements with some recent experiments [Sca04] but disagree with other experiments that have shown a correlation between the native structural properties and the folding rates [Pla98].

Notice that this kind of models do not present an intrinsic hierarchical structure (primary, secondary and tertiary) as in proteins. They should be rather thought of as a rough modelling of a mixture of secondary and tertiary contacts. This does not belittle the use of these studies since it is known that the secondary structures generally form before or meanwhile the tertiary structure does. Indeed, in general terms, it is believed that there are three kinds of possible folding mechanisms: i) the hierarchical mechanism where the secondary structures form before tertiary contacts, ii) the nucleation-condensation mechanism where a set of secondary contacts initiates the growth of the native state and iii) the hydrophobic collapse mechanism where tertiary hydrophobic contacts initiate the secondary structures. In all cases, a mix of secondary and tertiary structures precede the transition state and the precise folding mechanism may strongly depend on each specific case.

4.1.2. Designed heteropolymers

It is numerically possible to design heteropolymers, with non-covalent random interactions, that show a protein-like behaviour [Sha94, Sha93]. To this end, let us consider a heteropolymer on a cubic lattice whose sequence is composed by N monomers m_i,

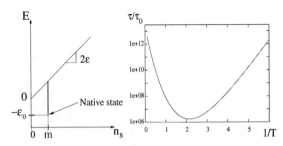

FIGURE 14. The modified Zwanzig model and the bell-shape curve of the folding time. The original Zwanzig model (see Fig. 4) can be modified (left picture) in order to take into account a native state shifted from the bottom of the valley $n_s = 0$. The corresponding folding time as a function of the temperature is reported on the right figure. In this figure $\varepsilon = 0.5$ and $m = 5$.

$i = 1...N$. The interaction energy E_{ij} of two adjacent and non-covalent monomers, m_i and m_j, is supposed to be a random quenched (i.e. fixed during all the procedure) variable with zero mean value and a variance 1 –this sets the energy unit. From the "residues" m_i and the matrix E_{ij}, the following design procedure leads to an heteropolymer that folds into a compact structure S. Given a sequence of the monomers defined by their position along the chain, one permutes two of them (which corresponds to an exchange mutation in an evolutionary terminology) and accepts the permutation if the total energy of the compact structure S decreases. At the end of this annealing procedure in the primary sequence space, one generally gets an heteropolymer that folds quickly [Sha94, Sha93]. Moreover, at sufficiently high temperature, two-state behaviour is often observed. The dynamics is usually a "coin and crankshaft" Monte-Carlo type with Metropolis acceptance rate, known to be ergodic [Hil75]. A hint to understand the propensity to fold is the presence of an energy gap between the native state and the lowest (in energy) misfolded states [Sha94]. However, the existence of a gap is not a sufficient condition for the molecule to fold since a flat energy landscape with a single local minimum energy state (i.e. a golf hole course) does not lead to a fast folder. As a consequence, it is reasonable to think that the annealing procedure in the primary sequence space indirectly designs a funnelled energy landscape and not only a single thermodynamically stable state.

In numerical studies, one has access at any time to the total number of native contacts, the number of native contacts of each monomer, the structural overlapping (that quantifies the matching of the relative position of distant monomers), the end-to-end distance and the gyration radius. Such an amount of information has led to a good understanding of such systems. For instance, it has been shown that the folding rates are correlated with the parameter $\sigma = |T_\theta - T_F|/T_\theta$ [Kli96], T_θ and T_F being respectively the Flory coil-to-globule transition and the melting temperatures. The latter determines the first-order transition between the denatured and the native states. It has also been shown that the size-dependence of the folding rates is sensitive to the degree of design [Gut96]. Resistance to mutations has also been studied [Bro99], the main result being that the latter directly depends on the magnitude of the energy gap. By further adding random interactions to E_{ij}, and by including hydrophobicity, the phase diagram in the temperature and denaturant concentration (related to the amount of extra disorder) phase has revealed the presence of a thermodynamic transition line between compact native structures and coil states but also between native and compact denatured states as suggested by experiments [Fin02]. The latter are good candidates to be intermediate states to the folding.

The bell-shape of the folding time. The above designed heteropolymers lead to a folding time that exhibits the bell-shape temperature dependence observed in experiments (Fig. 14) [Onu97]. The origin of this non-monotonic behaviour can be twofold. First, it can be due to the roughness of the energy landscape, which becomes the limiting rate factor when the thermal energy is on the order of the energy barriers separating the multiple configurations associated to the denatured state [Bry89, Onu97]. The simplest corresponding model describing this scenario is due to Zwanzig [Zwa95]. By introducing, in the microscopic time scale of the original model of Fig.4, a multiplicative Arrhenius factor $\exp(\Delta E/k_B T)$, one recovers a non-monotonic behaviour for the folding time. The argument is as follows. At high temperature (entropic regime), the folding time is large because of the high entropy of the denatured states (see Fig. 4). At low temperature, the folding time is large because of the trapping of misfolded states whose presence is reflected in the modified microscopic timescale. The life-time of the latter is on the order of $\exp(\Delta E/k_B T)$. Second, it can also be the manifestation of a crossover between a regime dominated by entropic effects (high temperature) and a regime dominated by activated events not related to any glass transition [Gut98]. As an example, let us consider the picture as proposed by Zwanzig [Zwa95]. Instead of taking into account a native state at the bottom of a potential energy valley, we can define a native state shifted with respect to the bottom of the valley. If the native state is shifted to a distance m (Fig. 14), a calculation that assumes partial equilibration out of the native state leads to a folding time $\tau = f(m)$ [Jun06] that has a bell-shape as reported in Fig. 14. In general, the folding time is given by $\tau = g(n_s^*)$ where n_s^* is the average distance between the unfolded state and the native state. The minimum folding time then corresponds to $n_s^* = m$. At high temperatures, the entropy favours large n_s^* whereas at low temperature, $n_s^* \approx 0$ dominates. In this case, the dynamics is activated and one finds an Arrhenius law $\tau \sim \exp(2m\varepsilon/T)$.

Force-induced transitions. A few numerical investigations of designed heteropolymer sequences in the presence of force have been done. The study in [Soc99] has revealed a tricky interplay between different reaction coordinates, e.g. the end-to-end distance and the number of native contacts. This reminds eventual problems in interpreting the diffusive dynamics in a projected free energy landscape.

More generally, interesting investigations by Geissler and Shakhnovich [Gei02] have shown that stretched designed heteropolymers should behave differently than stretched random heteropolymers. In particular, they argue that only protein-like sequences would reproducibly unfold and refold at a specific force. Also related to the protein-like behaviour, it has been shown that a simple stretched polymer at a temperature smaller than its θ-temperature (the coil-to-globule transition) leads at some force to the formation of the α-helix secondary structure [Mar03].

Stretched designed heteropolymers on-lattice can illustrate the presence of on and off pathway states. Indeed, in some conditions of temperature and force, a three-state behaviour can be observed [Jun06] –see Fig. 15. We then carried out the force-protocol

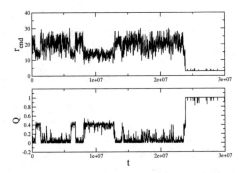

FIGURE 15. Three-state behaviour in a simulation of an heteropolymer on lattice. In this simulation, the mechanical force $\vec{f}$ is incorporated by adding a mechanical energy of the type $- \parallel \vec{f} \parallel \times \parallel \vec{r}_{end-to-end} \parallel$. This is different from the usual scalar product in order to prevent the geometrical effects of the lattice (details in [Jun06]). The upper panel shows the temporal evolution of the end-to-end distance and the lower panel shows the corresponding evolution of the percentage of native contacts.

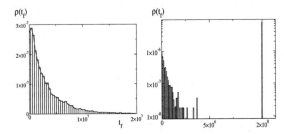

FIGURE 16. Left: the distribution of folding times from the intermediate states suggest a case where only on-pathway states are present. Right: in contrast, off-pathway-states are characterized by a peak at very large time (here 2×10^8). This peak actually corresponds to a cut-off in the simulation and would theoretically correspond to an infinite time. In this example, one finds 48% of on-pathway states and 52% of off-pathway states.

described in section 3.4 to quantify the fraction of misfolded states with respect to the on-pathway states. To this end, each time the system reaches an extension and a percentage of native contacts compatible with an intermediate state, the force is set to zero and the distribution of folding time from this very moment is computed. In small sized systems, fluctuations are important and an unfolding event at zero force is always observed. As a consequence, we numerically constrained the system to stay in the phase space region corresponding to the intermediate state (details in [Jun06]). In a situation with only on-pathway states, the distribution is nearly exponential as reported in Fig. 16(left). When off-pathway states are present, the distribution consists of two well-separate contributions (Fig. 16(right)). The peak observed at very large times is a numerical cut-off time after which we decided to stop the simulation. This peak corresponds to the off-pathway states.

4.2. Coarse-grained models

Coarse-grained simulations allow us to investigate folding kinetics up to thousands of microseconds. This is not possible by using standard molecular dynamics due to limited computing power. The underlying reason to deal with coarse-grained models is the belief that microscopic details are not determinant to understand the folding process [Bak00]. Usual coarse-grained procedures [Bak00, Guo95, Zho99] are inspired from simple Gō-like models such as the HP models. These involve an off-lattice dynamics of only the C^α carbons of the polypeptide chain. A typical model considers three types of carbons (or beads in the literature) that can be hydrophobic B, hydrophilic L or neutral N [Vei97]. The energies involved can be divided into two parts: local and non-local. The local contribution accounts for the covalent bonds and takes into account harmonic bonds and angle potentials, and a dihedral angle potential chosen to favour different orientations according to the surrounding secondary structure [Vei97, Zho99]. Non-local interactions count for the non-bonded interactions that are responsible for the folding mechanism. They are usually described by a Lennard-Jones potential and only interactions between hydrophobic pairs are taken into account. These models allow to study the statistical properties of the folding process starting from unfolded configurations, in contrast to the molecular dynamics simulation. Several results have been obtained about thermal folding [Bak00]. For instance, Zhou and Karplus [Zho99] have shown that a wide range of mechanisms could be observed in small helical proteins just by playing on the energy difference between the native and the non-native contacts.

In the spirit of the Gō-like models, let us also mention the use of mesoscopic elastic models which provide insight on protein dynamics and folding/unfolding pathways [Mic02]. At variance with other approaches, the strength of the non-covalent bonds depends on the temperature. This has led to identify some interesting differences with random heteropolymers, e.g. the structural regions involved in slow motions for protein-like models are much more extended than in random heteropolymers [Mic02].

In the presence of force, unfolding pathways seem to be mainly related to the structure of the native state [Kli00]. However, since the study of mechanical properties of proteins is still in its infancy [Ben96, Bus03], it would be rather audacious to say that one can in any case deduce the mechanical properties from the structure. Two major combined difficulties actually make this investigation difficult: 1) upon mechanical stretching, it is not clear how the network of forces is distributed inside the protein (for instance, three body interactions are numerous in the native state [Ejt04]), and 2) the mechanical properties at the single-residue level are not known.

The instructive RNA case. A seemingly simpler problem is the one of small RNA hairpins. Indeed, in good solvent conditions, the native structure corresponds to the secondary structure. The three dimensional structure is "only" constrained by the helix arrangement of the different base-pairs. Furthermore, the secondary structure is stabilized by stacking interactions whose values are well known [Jae93].

By adopting a similar coarse-grained model to the one described above, Hyeon and Thirumalai [Hye05] have studied in detail the differences between force and thermal induced unfolding. They have also studied the folding transition by using a force and a thermal jump protocols. Their study is extremely valuable since such protocols, es-

91

pecially the force jump experiment, have been realized in single-protein experiments [Fer04, Lee00] and in single RNA hairpins experiments as well [Pan06]. Their RNA coarse-grained description is composed of three beads that respectively correspond to the phosphate, the ribose and the base groups. A dihedral potential accounts for the right-handed chirality of RNA and a stacking stabilization potential is incorporated. Hydrophobic interactions between bases are described by a Lennard-Jones potential endowed with a distance cut-off and a Debye-Huckel electrostatic potential is introduced to describe the interaction between the phosphate groups. They use an overdamped Langevin dynamics and find, as in DNA experiments [War85], that thermal denaturation is due to the melting of the hairpin where each base-pair fluctuates independently. In contrast, mechanical denaturation occurs by sequential unzipping of the hairpin. An interesting prediction is that the refolding mechanism after a temperature quench should be different from that after a force quench. In particular, they find that the folding times upon force quench from stretched states, are much larger than those upon temperature quench from random states. They explain this phenomenon by the fact that stretched conditions make the molecule explore domains of phase space that are inaccessible at high temperatures by random coil configurations. Interestingly, such a statement is also valid for proteins and such reported differences are expected to occur in proteins as well.

5. CONCLUSION

The increasing number of single-protein experiments is providing insight on the inner details of the protein-folding problem. More generally, the combination of single-molecule techniques, with and without force, provide new quantitative results that can be rationalized with existing theories. In the long term these experimental results will be useful to better understand the basic mechanisms underlying many biological processes at the molecular and cellular level.

The combination of detailed numerical simulations and experiments has sharpened the theory underlying the propensity of proteins to fold. In particular, it has confirmed the funnel-like shape of the energy landscape without excluding well defined steps in the successive stages of the unfolding/folding transition. The confrontation of different experimental single-molecule techniques using mechanical force, on one hand, and fluorescence techniques, on the other hand, raises new interesting questions about the nature of the early intermediate states that form during the folding process. The following questions can now be answered with the recently available techniques: Under which conditions can mechanical force stabilize different intermediate states? Is the intermediate state observed under force the same as the early state that forms without force? In order to answer these questions, future design of experiments is needed to obtain structural information about the intermediate states, the transition states but also the denatured states.

An alternative approach is to consider protein-like models on-lattice that show some of these behaviours. It is then possible to investigate the different scenarios in a given protein and clarify whether the two intermediates, that are found with and without force, are the same or not. In this regard, the mechanical response studied at various forces, or by applying the force at different locations along the polypeptide chain, is expected to be different depending on which scenario is correct. Finally, further simulations of coarse-grained protein models, in conjunction with experimental measurements, might lead to improved models that faithfully reproduce the unfolding/folding pathways of proteins with and without force.

ACKNOWLEDGMENTS

We thank M. Palassini for useful comments. I. J acknowledges financial support form the European network STIPCO, Grant No. HPRNCT200200319. F. R acknowledges financial support from the Ministerio de Eduación y Ciencia (Grant FIS2004-3454 and NAN2004-09348) and the Catalan government (Distinció de la Generalitat 2001-2005, Grant SGR05-00688).

REFERENCES

All03. Allemand J. F., Bensimon D., Croquette V., Curr. Opin. Struct. Biol **13**, 266 (2003)
Anf73. Anfinsen C. B., Science **181**, 223Ũ230 (1973)
Bas97. Basché T., Moerner W. E., Orrit M. and Wild U. P. in *Single-molecule optical detection, imaging and spectroscopy*, WILEY-VCH, Weinheim, Cambridge (1997)
Bak00. Baker D., Nature **405**, 39 (2000)
Bal99. Baldwin R. L., Rose G. D., Trends Biochem. Sci. **24**, 77 (1999)
Ben96. Bensimon D., Structure **4**, 885 (1996)
Bla04. Blanchard S. C., Kim H. D., Gonzalez R. L., Jr., Puglisi J. D. and Chu S., Proc. Natl. Acad. Sci. USA **101**, 12893 (2004)
Boc02. Bockelmann U., Thomen P., Esevaz-Roulet B., Viasnoff V. and Heslot F., Biophys. J. **82**, 1537 (2002)
Bra03. Bradley P. *et al.*, Prot. Struct. Funct. Gen. **53**, 457 (2003)
Bro99. Broglia R. A., Tiana G., Roman H. E., Vigezzi E. and Shakhnovich E., Phys. Rev. Lett. **82**, 4727 (1999)
Bry89. Bryngelson J. D. and Wolynes P. G., J. Phys. Chem. **93**, 6902 (1989)
Bus03. C. Bustamante, Z. Bryant and S B. Smith, Nature **421**, 423 (2003) Special edition for the 50th anniversary of the discovery of DNA double helix. See references there.
Bus05. Bustamante C., Liphardt J., Ritort F., Physics Today **58** 43 (2005)
Cha89. Chan H.S, and Dill K. A., J. Chem. Phys., **90**, 492 (1989)
Cha94. Chan H.S, and Dill K. A., J. Chem. Phys., **100**, 9238 (1994) and references in there.
Cec05. Cecconi C., Shamk E. A., Bustamante C. and Marqusee S., Science **309**, 2057 (2005)
Cle03. Clementi C., Garcia A. E. and Onuchic J. N., J. Mol. Biol. **326**, 933 (2003)
Coc03. Cocco S., Marko J. F. and Monasson R., Eur. Phys. J. E **10**, 153 (2003)
Col05. Collin D., Ritort F., Jarzynski C., Smith S. B., Tinoco I. and Bustamante C., Nature **437**, 231 (2005)
Cor95. Cornell W. D. *et al.*, J. Am. Chem. Soc. **117**, 5179 (1995) and references in there.
Cre92. Creighton T., Proteins Structure and Molecular Properties, Freeman, New York (1992)
Dav94. Davidson A. and Sauer R., Proc. Natl. Acad. Sci. USA **91**, 2146 (1994)
Der80. Berrida B., Phys. Rev. Lett. **45**, 79 (1980)
Dil97. Dill K. A. And Chan H. S., Nat. Struc. Biol. **4**, 10 (1997)
Dob98. Dobson C. M., Sali A. and Karplus M., Angew. Chem. Int. Ed. Eng. **37**, 868 (1998)
Edi96. Ediger M. D., Angell C. A., Nagel S. R., J. Phys. Chem. **100**, 13200 (1996)
Ejt04. Ejtehadi M. R., Avall S. P. and Plotkin S. S., Proc. Natl. Acad. Sci. USA **101**, 15088 (2004)
Fer04. Fernandez J. M. and Li H., Science **303**, 1674 (2004)
Fis00. Fisher T. E., Piotr M. E. and Fernandez J. M., Nat. Struct. Biol. **7**, 719 (2000)

Fis99. Fisher T. E., Oberhauser A. F., Vezquez M. C., Marsalek P. E. and Fernandez J., Trends Biochem. Sci. **24**, 379 (1999).

Fit04. Fitzkee N. C. and Rose G. R., Proc. Natl. Acad. Sci. USA **101**, 12497 (2004)

Fin02. Finkelstein A. V., Ptitsyn O. B. in *Protein Physics*, Soft condensed mater complex fluids and biomaterial series, Academic Press (2002)

Gar02. Garcia-Mira M. M., Sadqi M., Fischer N., Sanchez-Ruiz J. M. and Muñoz, Science **298**, 2191 (2002)

Gei02. Geissler P. L. and Shakhnovich E. I., Phys. Rev. E **65**, 056110 (2002)

Gor06. Gore J., Bryant Z., Stone M. D., Nöllmann M., C. N. R., Bustamante C., Nature **439**, 100 (2006)

Guo95. Guo Z. and Thirumalai D., Biopolymers **36** 83-102 (1995)

Gut96. Gutin A., Abkevich V. and Shakhnovich E. I., Phys. Rev. Lett. **77**, 5433 (1996)

Gut98. Gutin A., Sali A., Abkevich V., Karplus M., and Shakhnovich E. I., J. Chem. Phys. **108**, 6466 (1998)

Ha. http://bio.physics.uiuc.edu/newTechnique.html, the Ha group in Urbana, Illinois

Ham55. Hammond, G. S., J. Am. Chem. Soc. **77**, 334 (1955)

Hil75. Hillorst H. J. and Deutch J. M., J. Chem. Phys. **63**, 5153 (1975)

Hye05. Hyeon C. and Thirumalai D., Proc. Natl. Acad. Sci. USA **102**, 6789 (2005)

Jae93. Jaeger A., Santa Lucia J., Jr., and I. Tinoco Jr., Annu. Rev. Biochem. **62**, 255 (1993)

Jun06. I. Junier and F. Ritort, in preparation

Kac06. Kachalo S., Hsiao-Mai L., Liang L., Phys. Rev. Lett. **96**, 058106 (2006)

Kar02. Karplus M. and McCammon J. A., Nat. Struct. Biol. **9**, 646 (2002) and references there.

Kau48. Kauzmann W., Chem. Rev. **43**, 219 (1948)

Kel97. Kellermayer M. S. F., Smith S. B., Granzier H. L. and Bustamante C., Science **276**, 1112 (1997)

Kli96. Klimov D. K. and Thirumalai D., Phys. Rev. Lett. **76**, 004070 (1996)

Kli00. Klimov D. K. and Thirumalai D., Proc. Natl. Acad. Sci. USA **97**, 7254 (2000)

Kuz05. Kuzmenkina E. V., Heyes C. D. and Nienhaus G. U., Proc. Natl. Acad. Sci. USA **102**, 15471 (2005)

Kuz06. Kuzmenkina E. V., Heyes C. D. and Nienhaus G. U., J. Mol. Biol. **357**, 313 (2006)

Lan03. Lang M. J. and Block S. M., Am. J. Phys. **71**, 201 (2003)

Laz97. Lazaridis T., Karplus M., Science **278**, 1928 (1997)

Lee00. Leeson D. T., Gaidagger F., Rodriguez H. M., Gregoret L. M. and Dyer R. B., Proc. Natl. Acad. Sci. USA **97**, 2527 (2000)

Lev68. Levinthal C., J. Chim. Phys. **65**, 44 (1968).

Mca99. McAdams H. H. and Arkin A., Trends Genet. **15**, 65Ũ69 (1999)

Man06. Manosas M., Collin D. and Ritort F., Phys. Rev. Lett., in press

Mar03. Marenduzzo D., Maritan A., Rosa A., Seno F., Phys. Rev. Lett., **90**, 088301 (2003)

Mic02. Micheletti C., Lattanzi G. and Maritan A., J. Mol. Biol. **321**, 909 (2002)

Mil02. Millett I. S., Doniach S. and Plaxco K. W., Advan. Protein Chem. **62**, 241 (2002)

Mun04. Muñoz V. and Sanchez-Ruiz J., Proc. Natl. Acad. Sci. USA **101**, 17646 (2004)

Onu04. Onuchic J. N. and Wolynes P. G., Curr. Op. Struct. Biol. **14**, 70 (2004)

Onu97. Onuchic J. N., Luthey-Schulten Z. and Wolynes P. G., Annu. Rev. Phys. Chem. **48**, 545 (1997)

Pan06. Pan L., Collin D., Smith S. B., Bustamante C. and Tinoco I., Jr., Biophys. J. **90**, 250 (2006)

Pan98. Pande V. S. and Rokhsar D. S., Proc. Nat. Acad. Sci. USA **95**, 1490 (1998)

Pet04. Petsko G. A. and Ringe D. in *Protein Structure and Function*, J. Heredity **95** 274 (2004)

Pla98. Plaxco K. W., Simons K. T. and Baker D., J. Mol. Biol. **277**, 985 (1998)

Ras99. Raschke T. M., Marqusee S., Nat. Struc. Biol. **6**, 825 (1999)

Rie97. Rief M., Gautel M., Oesterhelt F., Fernandez J. M. and Gaub H. E., Science **276**, 1109 (1997)

Ris89. Risken H. in *The Fokker-Planck Equation: Methods of Solution and Applications*, Springer-Verlag, Berlin (1989)ă

Rit03. Ritort F., Poincare Seminar, **2** 193 (2003)

Rit02. Ritort F., Bustamante C., Tinoco I., Proc. Natl. Acad. Sci. USA **99**, 13544 (2002)

Rit06. Ritort F., Mihardja S., Smith S. B. and Bustamante C., Phys. Rev. Lett. **96**, 118301 (2006)

Sca04. Scalley-Kim M. and Baker D., J. Mol. Biol. **338**, 573 (2004).

Sch05. Schlierf M. and Rief M., Biophys. J. **90**, L33 (2005)

Sha94. Shakhnovich E., Phys. Rev. Lett. **72**, 3907Ũ3910 (1994)

Sha93. Shakhnovich E. I. and Gutin A. M., Proc. Natl. Acad. Sci. USA **90**, 7195 (1993)

Smi01. Smith D. E. *et al.*, Nature **413**, 748 (2001)

Sno02. Snow C. D., Nguyen H., Pande V. S. and Gruebele M., Nature **420**, 102 (2002)

Soc99. Socci N. D., Onuchic J. N. and Wolynes P. G., Proc. Natl. Acad. Sci. USA **96**, 2031 (1999)

Str96. Strick T. R., Allemand J. F., Bensimon D., Bensimon A. and Croquette V, Science **271**, 1835 (1996).

Str00. Strick T. R., Croquette V. and Bensimon D., Nature **404**, 901 (2000).

Tan03. Tan E., Wilson T. J., Nahas M. K., Clegg R. M., Lilley D. M. J., and Ha T., Proc. Natl. Acad. Sci. USA **100**, 9308 (2003)

Tan66. Tanford C., Kawahara K. and Lapanje S., J. Biol. Chem. **241**, 1921 (1966)

Thi97. Thirumalai D., Klimov D. K., Woodson S. A., Theor. Chem. Acc. **96**, 14 (1997)

Thi05. Thirumalai D. and Changbong H., Biochem. **44**, 13 (2005)

Vei97. Veitshans T., Klimov D. K. and Thirumalai D., Folding Des. **2**, 1 (1997)

War85. Wartell R. M. and Benight A. S., Phys. Rep. **126**, 67 (1985)

Zho99. Zhou Y. and Karplus M., Nature **401**, 400 (1999)

Zwa92. Zwanzig R., Szabo A. and Bagchi B., Proc. Natl. Acad. Sci. USA **89**, 20 (1992)

Zwa95. Zwanzig R., Proc. Natl. Acad. Sci. USA **92**, 9801 (1995)

Zwa01. Zwanzig R., Nonequilibrium Statistical Mechanics, 1st Ed. (Oxford University Press, Chapter 4 (2001).

Active Sites by Computational Protein Design

Pablo Tortosa* and Alfonso Jaramillo*,†

*Laboratoire de Biochimie. CNRS - UMR 7654. Ecole Polytechnique, 91128 Palaiseau, France
†Biocomputation and Complex Systems Physics Institute (BIFI), University of Zaragoza, 50009 Zaragoza, Spain

Abstract. We have developed an automated method to design active sites into protein scaffolds using computational protein design techniques. We search through the amino acid sequence and conformation spaces by optimising protein stability and ligand binding. We use an all-atom force field, a high- resolution protein structure and a rotamer library to model a protein's unfolded and folded states. We enlarge a rotamer library by using a minimization procedure that optimizes rotamers to maximize intermolecular h-bonds. We validate our methodology by re-designing SH3-domain proteins to bind a set of 64 peptides.

Keywords: Computational protein design, rotamer libraries, molecular mechanics, combinatorial optimization
PACS: 83.10.Rs

INTRODUCTION

In the last few years there has been a tremendous advance in the use of automatic procedures in computational protein design. This is due to a better understanding and modelisation of the physical interactions of the atoms that form a protein, together with the reformulation of the problem as a combinatorial optimization issue. The group of Prof. Mayo (Caltech, USA) was the first to succeed in designing an artificial protein sequence able to fold into a predefined fold [1]. Starting from a high-resolution protein backbone structure, he used his combinatorial methodology to simultaneously mutate all residues in all possible ways. Recently, the group of Prof. Baker (Univ. of Washington, USA) used a more empirical methodology to design a protein having a new protein fold [2], opening the door to a protein-based nanotechnology. Later, the group of Prof. Hellinga (Duke Univ., USA) made another remarkable breakthrough [3]. They used their computational methods to introduce a catalytic activity into an inert protein scaffold by simultaneously mutating 20 residues. Here they reduced the problem of designing a catalytic site to the stabilization of the protein binding to a high-energy intermediate, similarly to the approach used to generate catalytic antibodies.

With the current computer facilities, more and more ambitious projects are being attacked. In the problem of protein-ligand recognition, large libraries are needed due to the combination of docking and side-chain placement requirements. As the number of pair-energies to be computed grows roughly as N_{rot}^2, the number of rotamers that can enter in a calculation is limited to time and memory requirements and some design problems are difficult to study with the required degree of refinement.

An usual approach is to refine the library in the docking region, but rotamer conformations from a standard library may not be appropriate for the specific conditions of

CP851, *From Physics to Biology; BIFI 2006 II International Congress,*
edited by J. Clemente-Gallardo, Y. Moreno, J. F. Sáenz Lorenzo, and A. Velázquez-Campoy
© 2006 American Institute of Physics 0-7354-0350-3/06/$23.00

the protein-ligand interface. Moreover, no libraries exist to optimize the binding to arbitrary ligands. Using sub-rotamers involved in recognition is an interesting possibility, but the number of possible interacting pairs is so large that a proper minimization for each pair is too time consuming. Here we present a fast approach that refines the rotamer library by generating new rotamers that have at least an h-bond with the ligand. We also apply the refining step to pairs of rotamers interacting with the ligand. We expect that the satisfaction of a maximal number of intermolecular h-bonds together with shape complementarity will help to achieve specific binding. We test the methodology by re-designing SH3-domain proteins to bind 64 peptides.

METHODS

Our computational protein design software DESIGNER [5] has already been successfully tested by redesigning seven folds [5] and by designing peptides that bind to MHC-I proteins [6]. DESIGNER uses a high-resolution atomic structure of a protein as scaffold, (either from x-ray determinations or obtained by a structure model if no x-ray determination exists), a standard molecular mechanics force field (CHARMM22 [7]) and an implicit solvation model based on atomic surface areas with experimental atomic hydration coefficients obtained from small molecules [8]. A local rotamer library is added at each residue position to generate the atomic coordinates of each possible mutated residue. Either the complete set or a subset of sidechains are allowed to vary randomly while keeping a fixed backbone.

The score used to rank the mutant sequences is an approximation to the folding free energy, which is computed by modelling both an unfolded and folded state at the atomic level with the electrostatics, van der Waals and implicit solvation terms. In our approximation, the folding free energy of any possible sequence can be written in terms of pairwise backbone-backbone (including main-chain and fixed, non-designed residues), single (between rotamer and backbone) and pair energies (between each pair of interacting rotamers):

$$E(x) = E_{fixed} + \sum_i E_{singles}(x_i) + \sum_{i<j} E_{pairs}(x_i, x_j) \qquad (1)$$

In this formulation, all interactions can be precomputed resulting in an enormous saving of computing time in the optimization step.

Original rotamer library construction. The rotamers from a backbone-dependent library [4] are associated to each position. A rotamer-backbone optimization of 100 steps of conjugate gradient is performed in the context of a fixed backbone devoided of all designable residues. This has the effect of reducing potential clashes and allowing for favorable interactions not included in the original library, adapting the rotamers to the local backbone conditions. The resulting energy is stored as the single energy of the rotamer (see Eq. (1)). In each position, rotamers with an energy difference with respect to the minimal energy rotamer above a given threshold are eliminated to reduce the size of the combinatorial space. Initial rotamer coordinates are updated with those resulting from this optimization step and are kept for the rest of the calculation.

Pair energies calculation. In order to avoid possible clashes and allow for favorable interactions, we perform a pair-rotamer optimization: each pair of interacting rotamers is minimised with 8 steps of steepest descent and the interaction energy is stored as the pair energy (see Eq. (1)). We use the original van der Waals radii from the CHARMM22 force-field.

Side-chain optimization. Optimization techniques are finally needed to find the set of possible sequences stabilising the corresponding fold. The size of the optimization problem makes exact solving methods difficult to apply, and combinatorial optimization techniques are preferred. For the side-chain modeling calculations we use a simulated annealing optimization with an initial temperature of $T_0/RT = 50$ and 5000 iterations. We repeated the procedure several times to assure convergence.

H-bond optimized rotamer library

We have used the following protocol to refine the library only for the interface positions, searching for conformations that provide a better interaction and increase the number of h-bonds. For each pair of rotamers involved in binding we first minimize the structure (100 iterations of steepest descent) of the interacting polar rotamers. A vacuum environment in CHARMM ($\varepsilon = 1$) was found to significantly speed-up calculations by biasing the search toward h-bond formation. Backbone charges are set to zero to restrict the optimization to rotamer-rotamer interactions. In a second step, we minimize ($\varepsilon = 8$) the resulting structure with 100 steps of Adopted Basis Newton-Raphson (ABNR) to obtain the refined conformation of both residues, which are stored as new rotamers for further use. This produced a total of 20938 rotamers. As a last step, rotamers differing in less than 2 degrees in dihedral angle were removed, leaving a total of 14734 rotamers (29.6% reduction).

RESULTS

We have considered the N-terminus SH3 domain of the CRK protein for which a high resolution structure is available complexed with the proline-rich peptide PPPALPPKKR (PDB code 1cka, 1.4 Å resolution). We have simultaneously mutated all protein positions (except positions containing prolines) that had an atom less than 10 Å away from the peptide in the x-ray structure. We have allowed to mutate to any amino acid except proline. For the peptide, we have mutated positions 5, 8 and 9 with a reduced alphabet of charged atoms (D, E, R and K) while keeping position 10 as wild type. This gives 31 positions and a total of 4453 rotamers, with a size of the combinatorial space of 10^{67} possible structures. The improved rotamer library consists of 17208 rotamers, with a size of the combinatorial space of 10^{78}. To assess the ability of the optimized rotamer library to improve design results, we have used a reduced alphabet (D,E,R,K) for peptide positions 5, 8, and 9 and redesigned the SH3-domain protein for specific binding to each of the 64 possible peptides.

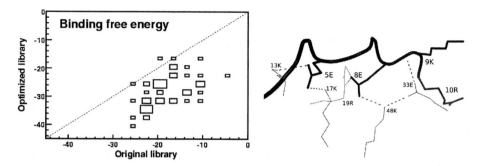

FIGURE 1. Left) Improvement of the binding free energy between the peptide and the designed protein when using the optimized rotamer library. The size of each box in the 2D histogram is proportional to the population of the bin. The average improvement is -10.3 kcal/mol. Right) Best-binding design obtained with the optimized rotamer library (peptide PPPAEPPEKR) with most relevant interactions.

We have selected the best 10 peptide-protein complexes with the h-bond optimized rotamer library and compared to the results obtained independently using only our original library[1] (composed by Dunbrack's backbone dependent library [4], rotamers with $P > 90\%$). Figure 1 shows an average improvement of -10.3 kcal/mol for each peptide, which corresponds to -7.6 kcal/mol for the folding free energy (data not shown). Table 1 shows in the left the folding free energy and binding free energy of the best 10 peptides (with strongest binding) obtained by designing both the protein sequence and the peptide sequence, using the h-bond optimized library. At the right we compare those results with those obtained with the original library, both for i) the same protein and peptide sequence used in the left and ii) the fully designed protein and peptide. Notice that the case i) would correspond to perform side-chain modelling on the protein and peptide sequences obtained in the design using the optimized library. In the other hand, ii) corresponds to do a design of the protein sequence but keeping constant the peptide sequence (allowing this one to change side-chain conformation). The average improvement obtained is -7.6 kcal for fixed sequences. Only 6 of the 10 best designed binders found with the h-bond optimized library are also in the ten best binders list obtained when the protein is designed using only the initial library.

DISCUSSION

We have developed a methodology able to combine the exploration of sequence-rotamer space with the problem of designing a specific binding site in to a single combinatorial optimization problem. This has the advantage of allowing the exploration of a much larger number of combinations than previous methods that sequentially perform several cycles of protein design followed by active site design. The key element was the design

---

[1] The rotamers from the optimized library include as a subset those in the original library

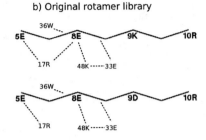

FIGURE 2. Schematic view of the binding interaction network of two of the best-binding sequences (1st, PPPAEPPEKR, and 6th, PPPAEPPEDR) for a) the optimized version and b) the original rotamer library.

TABLE 1. The left hand side shows the energies corresponding to the 10 best-binding sequences obtained by designing the protein and peptide aminoacid sequences using the h-bond optimized library. The right hand side shows the corresponding results obtained with the original library for: i) a side-chain modelling using the same protein and peptide sequence than used in the left and ii) a full design of the protein (but keeping fixed the peptide sequence) using the original library. Only positions 5, 8 and 9 of the peptide were designed. The total folding energy of the complex and the contributions of electrostatics + van der Waals and solvation to binding is shown. The number of h-bonds (involving peptide side-chains) and the order of each peptide's correspondent design in the 10 best binders list with the original library is also shown.

	Optimized library					**Original library**					
						Same seq.		Designed			
	E (kcal/mol)					E (kcal/mol)		E (kcal/mol)			
Peptide	Total	Bind	e+vW	sv.	hb	Total	Bind	Total	Bind	hb	#
PPPAEPPEKR	-181.3	-40.5	-49.8	8.5	6	-167.4	-34.2	-173.0	-26.9	3	1st
PPPADPPKKR	-181.5	-37.5	-41.0	3.5	3	-169.8	-31.9	-175.3	-24.1	2	4th
PPPADPPKRR	-181.0	-37.3	-41.1	3.8	3	-171.0	-32.8	-176.2	-22.1	3	
PPPADPPKER	-181.9	-36.2	-37.6	1.4	3	-171.0	-31.6	-177.1	-23.8	4	7th
PPPAEPPKKR	-182.9	-35.9	-39.6	3.7	3	-170.4	-31.4	-175.8	-23.2	3	9th
PPPAEPPEDR	-178.0	-35.7	-44.7	9.0	7	-164.7	-26.1	-170.4	-23.9	2	5th
PPPADPPDKR	-180.1	-35.4	-40.2	4.8	2	-160.8	-22.4	-167.2	-21.7	3	
PPPADPPDRR	-179.6	-34.9	-30.3	5.4	4	-161.9	-23.3	-168.6	-21.8	4	
PPPAEPPKER	-183.2	-34.6	-36.2	1.6	5	-171.7	-31.1	-177.6	-22.9	5	10th
PPPADPPDER	-180.4	-34.5	-37.0	2.5	3	-162.1	-22.0	-168.9	-21.2	5	

of good binders with a high specificity, which we assumed was possible by a maximal satisfaction of intermolecular h-bonds. This was achieved by the use of pairwise minimization cycles to explore the conformational degrees of freedom in such a way that our rotamer formed h-bonds with neighboring side chains. We stored those conformations that were able to provide h-bonds as new additional rotamers. We repeated this procedure including three-body contacts. Then, our enlarged library contained new rotamers able to provide binding modes with a high number of satisfied h-bonds. As the previous rotamers are sufficient to provide appropriate shape complementarity, we have reduced the problem of docking to a generalized rotamer search. We have applied our computational protein design methodology to design a set of 64 artificial SH3-domain proteins able to bind to 64 different ligands taken from mutating the WT ligand, a proline rich peptide. In this way we have produced an artificial sampling of different binding modes due to the redesign of the binding interface. To reduce the complexity of the problem and improve sampling, we have considered a fixed backbone for our ligand.

We could see in our results that we have obtained stronger binders using our methodology, compared with standard computational protein design. Although we cannot assure specificity, it is likely that we could get specific binding if we design binding modes able to satisfy all possible intermolecular h-bonds.

Compared to other computational approaches based on partially knowledge-based potentials, our approach is able to scan a larger conformational search and uses a general physicochemical rationalisation. As a clear advantage, our approach uses the same software and set of parameters to design new proteins for protein-DNA, protein-protein and protein-ligand interactions.

ACKNOWLEDGMENTS

A.J. acknowledges the financial support from HPC-Europa (FP6 EU project RII3-CT-2003-506079), from the "Programa de Incentivo a la Investigación (Universidad Politécnica de Valencia)" and the use of the CESCA and IDRIS supercomputer facilities to perform the calculations reported here. P.T. acknowledges the financial support by a long-term EMBO fellowship.

REFERENCES

1. B.I. Dahiyat and S.L. Mayo. De novo protein design: fully automated sequence selection. *Science* **278**, 82-87 (1997).
2. B. Kuhlman, G. Dantas, G.C. Ireton, G. Varani, B.L. Stoddard, D. Baker. Design of a novel globular protein fold with atomic-level accuracy. *Science* **302**, 1364-1368 (2003).
3. M.A. Dwyer, L.L. Looger, H.W. Hellinga. Computational Design of a Biologaically Active Enzyme. *Science* **304**, 1967-1971 (2004).
4. Dunbrack, R. L., Jr. & Karplus. Backbone-dependent rotamer library for proteins. Application to side-hain prediction. *J. Mol. Biol.* **230**, 543-574 (1993).
5. A. Jaramillo, L. Wernisch, S. Hery and S.J. Wodak. Folding free energy function selects native-like protein sequences in the core but not on the surface. *Proc. Natl. Acad. Sci.* **99**, 13554-13559 (2002).
6. K. Ogata, A. Jaramillo, W. Cohen, J. Briand, F. Connan and S.J. Wodak. Automatic Sequence Design of MHC Class-I Binding Peptides. Impairing CD8+T Cell Recognition. *J. Biol. Chem.* **278**, 1281-1290 (2003).
7. MacKerell, A.D., Jr., Bashford, D., Bellott, M., Dunbrack, R.L., Jr., Evanseck, J. D., Field, M. J., Fischer, S., Gao, J., Guo, H., Ha, S., Joseph-McCarthy, D., Kuchnir, L., Kuczera, K., Lau, F. T. K., Mattos, C., Michnick, S., Ngo, T., Nguyen, D. T., Prodhom, B., Reiher, W.E., III, Roux, B., Schlenkrich, M., Smith, J. C., Stote, R., Straub, J., Watanabe, M., Wiorkiewicz-Kuczera, J., Yin, D. & Karplus, M.. All-atom Empirical Potential for Molecular Modeling and Dynamics Studies of Proteins.. *J. Phys. Chem* **102**, 3586-3616 (1998).
8. Ooi, T., Oobatake, M., Nemethy, G., Scheraga, H.A.. Accessible surface areas as a measure of the thermodynamic parameters of hydration of peptides. *Chemistry* **84**, 3086-3090 (1987).

The Effects of H-Bond Cooperativity upon the Secondary Structures of Peptides

J. J. Dannenberg

Dept. of Chemistry and Biochemistry, City University of New York - Hunter College and The Graduate School, 695 Park Ave., New York NY 10021, USA

Abstract. Molecular orbital calculations show that amidic H-bonds within peptide structures can be extremely cooperative in some cases (α-helices), or not (collagen-like triple helices), while in others (β-sheets) cooperativity becomes masked by interactions that are weakened when new H-bonds are formed. There are cases where cooperativity is unimportant and H-bonding interactions can reasonably be approximated as being additive. In such cases, empirically derived pair-wise additive potentials might be expected to reproduce structures. However, those cases where cooperativity is important require more sophisticated analysis as the cooperative interactions between the individual H-bonds can become the dominant energetic driving force for the observed peptide structures. As in the case of many molecular crystals, pairwise additivity of the interactions of individual H-bonds can seriously underestimate the stability of the structure(s), leading to erroneous predicted structures.

The amidic H-bond (N-H...O=C) is ubiquitous in natural materials such as proteins, peptides and DNA. Nature uses the specific complementarity of the base pairs to determine the sequences of DNA : that of AT contain two H-bonds, while that of GC contains three. The structures of DNA's with different base sequences are virtually identical. In contrast, proteins and peptides have extremely varied structures, yet they are virtually all made from the same building blocks: the 20 naturally occurring amino acids. Each of these contains the same amide group with the same amidic H-bonding donors and acceptors except for proline which lacks an H-bonding donor. How does nature use similar H-bonding systems to regulate the structures of DNA and peptides, yet keep the structures of DNA quite regular while admitting an enormous structural heterogeneity to peptides? Of course, other factors (disulfide bridges, etc.) play important roles in peptides structures, but that of H-bonds can be extremely influential if not completely determining in some cases. We shall see that nature has taken advantage of the fact that these N-H...O=C H-bonds can vary in energy from a low of about 2[1] to at least 23[2] kcal/mol depending upon the extent of cooperativity between the H-bonds and the polarizability of the H-bonding molecules.

The results that follow are based upon molecular orbital (MO) calculations that used the B3LYP/D95(d,p) hybrid density functional theory (DFT) method often as part of an ONIOM[3] calculation that divides the system into two parts that use either high or low level calculations. The DFT method was used for the high and the AM1 semiempirical

CP851, *From Physics to Biology; BIFI 2006 II International Congress,*
edited by J. Clemente-Gallardo, Y. Moreno, J. F. Sáenz Lorenzo, and A. Velázquez-Campoy
© 2006 American Institute of Physics 0-7354-0350-3/06/$23.00

MO method for the low level. The H-bonding systems were always treated at the high level. Vibrational analyses were performed to obtain enthalpies of interaction (when reported as enthalpies) in many (but not all) cases. Geometric optimization was either calculated on counterpoise[4] corrected surfaces[5] or the single-point counterpoise correction was added to correct for basis set superposition error. More detailed descriptions can be found in the original reports that are referenced within.

Several years ago, we showed that H-bonding chains of formamides form highly cooperative H-bonds.[6,7] Thus, the calculated H-bonding enthalpy of a simple formamide dimer, 4.5 kcal/mol, increases by almost 200% to 13.0 kcal/mol when the H-bond is near the center of a long chain of 15 H-bonding formamides. This extraordinary cooperative effect cannot be explained simply by electrostatic interactions. If we consider the interactions between fixed dipoles aligned in a regular linear fashion with equivalent distances, r, between adjacent molecules, we must sum over all the pair-wise interactions between them. The interaction energies are simply the dipole moment times the applied dipolar electric field. Since the electric field of a dipole decreases as the third power of the distance from the center of the dipole, the field experienced by a second molecule due to the dipole of the first will be $(nr)^{-3}$ where n represents the number of intermolecular distances between the two molecules (in a liquid, the tumbling of the molecules with respect to each other reduces the dipole-dipole interactions to $(nr)^{-6}$ rendering cooperativity much less important). Thus, if we take the 1-2 (or nearest neighbor with n=1)) interaction to be some value, say K kcal/mol, then the 1-3 interaction between aligned dipoles will be $K/2^3$ or K/8. If we sum over all pairwise interaction in a chain of 15 dipoles, we would obtain a cooperative interaction of about 50% for the centermost H-bond strength, which would correspond to about 6.8 kcal/mol, far from the 13.0 observed.[6]

More recently, we have shown that 4-pyridone, which one can imagine as a formamide in which two parallel -C=C- are inserted between the C=O and the N-H forms much stronger H-bonds (9.9 kcal/mol in a dimer and up to 23 kcal/mol for the centermost H-bond in a long chain).[2] We attributed both the increased H-bonding strength and the increased cooperativity to the enhanced polarizabilty of 4-pyridone over formamide. The observation that the difference in the H-bonding energies hardly changes upon rotation

formamide   4-pyridone

Figure 1. Formamide can be transformed into 4-pyridone by rupture of the C-N bond and insertion of two parallel -C==C- units.

of adjacent pyridone rings about the H-bond from parallel to perpendicular positions either in the dimer or in long chains strongly suggests the role of covalent π-interactions to be negligible. However, σ-covalent interaction remain a possibility.

If we examine the relative energies of the secondary structures of peptides as a function of the kinds of amidic H-bonds present, we find that cooperativity (or lack of it) can determine the overall secondary structural motifs. In order to consider the relative energies of different structural motifs, we need some energy standard for comparison. We have chosen two different standards, as each is somewhat arbitrary. The first is a single completely extended strand of the peptide that contains no traditional H-bonds (as we shall see, C_5 H-bonding interactions persist and play a hitherto unsuspected role).[1,8] The second standard is the aggregate of the individual amino acids in their gas phase relaxed conformations. The energy of any peptide can be imagined as a polycondensation reaction of the requisite component amino acids (plus the necessary capping groups, if any) to form the peptide in question plus one water molecule for each amidic linkage formed.

Let us first consider two helical structures, the 3_{10}- and more common α-helices. In order to do so, we considered capped polyalanines, acetyl(ala)$_N$NH$_2$, where N varies between 2 and 18.[9] The geometries of each helical form and the standard extended strand were each optimized without any constraints using the ONIOM procedure. Stable optimized structures were found for 3_{10}-helices where N is 3 or greater. However, we only found stable α-helices for N=8, 10 and 12 or greater. In each of the helices, we found the H-bonds near the center to be shorter than those near the termini. Since H-bond energies generally correlate inversely with H-bond length, the centermost bonds are stronger by implication. The 3_{10}-helices contain two chains of amidic H-bonds (similar to the aforementioned formamide chains), while the α-helices contain three. The 3_{10}-helices were found to be generally more stable than the α-helical conformers, although the difference diminishes as N increases. For N=18, the difference all but disappears. Most tellingly, the increase in stability (relative to the extended strand) upon addition of one alanine to a growing chain exceeds 6 kcal/mol for α-helices for N >15, but has reached an asymptotic limit of about 3 kcal/mol at about N=12 for 3_{10}-helices. Clearly, the H-bond cooperativity is more significant for α- than 3_{10}-helices or extended strands. Our calculations agree with reports on somewhat similar systems by several other groups.[10] Kemp has recently shown that the helicity number, W, for Ala increases from about 1 ro 1.6 as the number as alanines increases.[11] He had previously noted the presence of enthalpic cooperativity in α-helix formation.[12]

Our calculations are supported by the experimental observations of Baldwin[13] and Kemp,[12,14] both of whom found that α-helices are favored for polyalanines in aqueous solution that have a minimum number of alanine residues similar to those that we found to be necessary. Starting from out completely optimized helical and extended strand geometries, we performed complete vibrational analyses. These lead to calculation of enthalpies (298 K) instead of energies simply determined by the differences in the local minima on the potential energy surfaces. Remarkably, our calculated stabilization enthalpies per alanine residue for α-helices versus extended strands (about 0.8 kcal/mol) are similar to those reported by Baldwin[15] (0.9).

We have calculated the effect of single mutations of the 10th Ala residue to Gly, Pro, Leu, Val, Phe and Ser in the capped polyalanine α-helix. The results for Pro and Gly are particularly instructive. Both favor the extend strand over the helix, but for very different reasons. Gly stabilizes both structures, but stabilizes the extended strand more, while Pro destabilizes both, but destabilizes the helix more.[16]

In contrast to α-helices, normal peptide (formed form α-amino acids) β-sheets have not generally been found to have cooperative H-bonding interactions despite the fact that the same kinds of H-bonding amide chain appear in these structures. On the other hand,

Small ring   Large ring

ΔH: -4.85 kcal/mol   -13.99 kcal/mol

Difference in ΔH = ~9 kcal/mol

Figure 2. The diglycine model strand in the center can dimerize to form either the small (left) of large (right) ring dimer. The cyclic cooperativity that couples the C_5 *intra*strand H-bonds with the *inter*strand ones causes the difference in H-bonding enthalpies

similar structures formed from β-amino acids do exhibit such cooperativity. Simple models of β-sheets formed from diglycine and tetraglycine models revealed the reason for this. A comparison of the two isomeric H-bonding dimers of our diglycine model is given in figure 2. The structure on the left contains a 'small ring' cyclic H-bonding motif, while that on the right has the 'large ring' motif. The difference in the H-bonding enthalpies of 9 kcal/mol surprised us (the H-bond energy of 2.4 kcal/mol per H-bond for the small ring structure represents the weakest N-H...O=C H-bond that we know of). The large ring structure contains a cyclic cooperative H-bonding structure that includes the *inter*strand H-bonds, as well as, the *intra*strand C_5 H-bonds (which become shorter upon formation of the dimer). On the other hand, the small ring structure in the left sacrifices the *intra*strand H-bonds (the O...H distance increases) to make the *inter*strand ones. The difference in H-bonding energies all but disappears if one inserts an extra -CH_2CH_2- spacer between the amide moieties in these two structures. The 9 kcal/mole difference between a large and small ring persists for all eight β-sheet models that contained two or four strands each containing two or four glycines. As a result of the formation of (roughly) equal numbers of small and large rings in antiparallel β-sheets, no significant cooperativity is found for adding more strands to the sheet.[1] However, we found cooperativity for the sheets formed with the $-CH_2CH_2-$ spacer,[1,17] and Wu found cooperativity for sheets made from β-amino acids.[18] A recent statistical survey of H-bond lengths from the protein data bank supports the suggestion that the H-bonds in α-

Optimized Triple Helix ProProGly

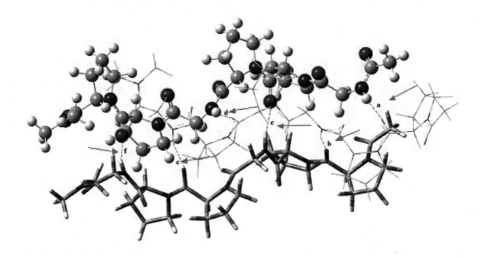

Figure 3. A collagen-like triple helix with the ProProGly repeating unit. Each of the three strands is indicated using a different kind of drawing (wireframe, ball and stick or tube) while the H-bonds ar indicated by the arrows.

helices are cooperative while those in β-sheets are not.[19]

We have applied similar methods to the study of collagen.[20] This structural protein forms triple helical structures from three peptide stands each of which has a Gly in every third residue. The other two positions in the strands can (in principle) contain any amino acids, however, Pro and hydroxyproline (Hyp) are particularly prevalent. We have found that the ONIOM calculation give H-bond lengths that agree well with crystallographic structures. Gly is the only achiral or enantiomorphic naturally occurring amino acid. We noted that the Gly residues behave as ᴰamino acids in collagen, using the N-H donor which would normally be the side chain in other ᴸamino acids to H-bond. This suggested that Gly might be replaced by ᴰamino acids without destabilizing the triple helix. In fact we caclulated that both ᴰAla and ᴰSer stabilize the triple helix, while ᴸAla destabilizes it (as was already known[21]). Since there are no chains of H-bonds in collagen, one would not expect H-bond cooperativity in these triple helical structures.

CONCLUSIONS

The examples discussed above suggest that the H-bond contribution to peptide (and by implication, protein) structure can be considered from several perspectives, we have discussed at length in a recent chapter:[22]

a) The liquid paradigm considers only nearest-neighbor interactions. This can be appropriate where cooperativity is not important. As we have seen above, 1-3 dipole-dipole interactions are reasonably neglected in the liquid state due to the r^{-6} relationship with distance.

b) The solid paradigm takes cooperativity (thus, non-nearest neighbor interactions) into account, as it must for structures with fixed relative orientations.

c) Nature has chosen the N-H...O=C H-bond as the H-bonding interaction of choice as its H-bonding interaction energies can be modulated as a function of the number of H-bonds in a contiguous chain (cooperativity) and the polarizability of the chemical linker between the N-H and C=O. Thus, the NH and C=O are directly bound to each other in H-bonding chains such as those in helices, while the polarizability of the connector in the large ring of β-sheets is increased by the C_5 H-bonds (see figure 2). The polarizability is further increased in 4-pyridone (which is a component structure of several purine or pyrimidine bases).

REFERENCES

1. R. Viswanathan, A. Asensio, and J. J. Dannenberg, *J. Phys. Chem. A* **108**, 9205 (2004).
2. Y.-f. Chen and J. J. Dannenberg, *J. Am. Chem. Soc.*, **128**, in press (2006).
3. M. Svensson, S. Humbel, R. D. J. Froese, T. Matsubara, S. Sieber, and K. Morokuma, *Journal of Physical Chemistry* **100**, 19357 (1996); S. Dapprich, I. Komiromi, K. S. Byun, K. Morokuma, and M. J. Frisch, *Theochem*, 0166 (1999).
4. H. B. Jansen and P. Ros, *Chem. Phys. Lett.* **3**, 140 (1969); S. F. Boys and F. Bernardi, *Mol. Phys.* **19**, 553 (1970).
5. S. Simon, M. Duran, and J. J. Dannenberg, *J. Chem. Phys.* **105**, 11024 (1996).
6. N. Kobko and J. J. Dannenberg, *J. Phys. Chem. A* **107**, 10389 (2003).
7. N. Kobko, L. Paraskevas, E. del Rio, and J. J. Dannenberg, *J. Am. Chem. Soc.* **123**, 4348 (2001).
8. V. Horvath, Z. Varga, and A. Kovacs, *J. Phys. Chem. A* **108**, 6869 (2004).
9. R. Wieczorek and J. J. Dannenberg, *J. Am. Chem. Soc.* **126**, 14198 (2004).
10. I. A. Topol, S. K. Burt, E. Deretey, T.-H. Tang, A. Perczel, A. Rashin, and I. G. Csizmadia, *J. Am. Chem. Soc.* **123**, 6054 (2001); Y.-D. Wu and Y.-L. Zhao, *J. Am. Chem. Soc.* **123**, 5313 (2001).
11. R. J. Kennedy, S. M. Walker, and D. S. Kemp, *J. Am. Chem. Soc.* **127**, 16961 (2005).
12. R. J. Kennedy, K.-Y. Tsang, and D. S. Kemp, *J. Am. Chem. Soc.* **124**, 934 (2002).
13. S. Marqusee, V. H. Robbins, and R. L. Baldwin, *Proc. Nat. Acad. Sci. U. S. A.* **86**, 5286 (1989).
14. J. S. Miller, R. J. Kennedy, and D. S. Kemp, *J. Am. Chem. Soc.* **124**, 945 (2002).
15. M. M. Lopez, D.-H. Chin, R. L. Baldwin, and G. I. Makhatadze, *Proc. Nat. Acad. Sci. U. S. A.* **99**, 1298 (2002).
16. R. Wieczorek and J. J. Dannenberg, *J. Am. Chem. Soc.* **127**, 17216 (2005).
17. Y.-L. Zhao and Y.-D. Wu, *J. Am. Chem. Soc.* **124**, 1570 (2002).
18. J.-Q. Lin, S.-W. Luo, and Y. D. Wu, *J. Comput. Chem.* **23**, 1551 (2002).
19. O. Koch, M. Bocola, and G. Klebe, *Proteins: Structure, Function, and Bioinformatics* **6**, 310 (2005).
20. M. I.-H. Tsai, Y. Xu, and J. J. Dannenberg, *J. Am. Chem. Soc.* **127**, 14130 (2005).
21. M. Bhate, X. Wang, J. Baum, and B. Brodsky, *Biochemistry* **41**, 6539 (2002).
22. J. J. Dannenberg, *Advances in Protein Chemistry* **72**, 227-73 (2006).

Effects of constraints in general branched molecules: A quantitative ab initio study in HCO-L-Ala-NH$_2$

Pablo Echenique[*,†], J. L. Alonso[*,†] and Iván Calvo[*,†]

[*]*Departamento de Física Teórica, Facultad de Ciencias, Universidad de Zaragoza, Pedro Cerbuna 12, 50009, Zaragoza, Spain.*
[†]*Instituto de Biocomputación y Física de los Sistemas Complejos (BIFI), Edificio Cervantes, Corona de Aragón 42, 50009, Zaragoza, Spain.*

Abstract. A general approach to the design of accurate classical potentials for protein folding is described. It includes the introduction of a meaningful statistical measure of the differences between approximations of the same potential energy, the definition of a set of Systematic and Approximately Separable and Modular Internal Coordinates (SASMIC), much convenient for the simulation of general branched molecules, and the imposition of constraints on the most rapidly oscillating degrees of freedom. All these tools are used to study the effects of constraints in the Conformational Equilibrium Distribution (CED) of the model dipeptide HCO-L-Ala-NH$_2$. We use ab initio Quantum Mechanics calculations including electron correlation at the MP2 level to describe the system, and we measure the conformational dependence of the correcting terms to the naive CED based in the Potential Energy Surface (PES) without any simplifying assumption. These terms are related to mass-metric tensors determinants and also occur in the Fixman's compensating potential. We show that some of the corrections are non-negligible if one is interested in the whole Ramachandran space. On the other hand, if only the energetically lower region, containing the principal secondary structure elements, is assumed to be relevant, then, all correcting terms may be neglected up to peptides of considerable length. This is the first time, as far as we know, that the analysis of the conformational dependence of these correcting terms is performed in a relevant biomolecule with a realistic potential energy function.

Keywords: constraints, protein folding, ab initio, conformational equilibrium, dipeptide, mass metric tensor, Fixman potential
PACS: 02.50.Cw, 31.15.Ar, 87.14.Ee, 87.15.Aa, 87.15.Cc

INTRODUCTION

Proteins are long chains comprised of twenty different amino acidic monomers and they are central elements in the biological machinery of all known living beings. They perform most of the catalytic tasks that are vital in the many coupled chains of chemical reactions occurring in the cells, they are found as structural building blocks in the cytoskeleton or in organelles, such as the ribosome, and they also play a very important role as membrane receptors. Their absence or malfunctioning is related to many diseases such as Creutzfeldt-Jakob's or Cancer [1, 2] and the proteins involved in the biology of pathogens are often the preferred target of newly designed drugs (see the contribution by C. Cavasotto in these proceedings).

Despite their complexity and the many opposing forces that determine their behaviour, these molecules swiftly acquire a unique three-dimensional *native* structure in the phys-

CP851, *From Physics to Biology; BIFI 2006 II International Congress,*
edited by J. Clemente-Gallardo, Y. Moreno, J. F. Sáenz Lorenzo, and A. Velázquez-Campoy
© 2006 American Institute of Physics 0-7354-0350-3/06/$23.00

iological milieu. Some details of this process are still not clear (see the contribution by J. M. Sánchez-Ruiz), such as the relative proportion of the naturally occurring proteins that fold co- or post-translationally (i.e., during or after biosynthesis at the ribosome) [3], or the role played by molecular chaperones such as GroEL, the Prolyl-peptidyl-isomerase, or the Protein disulfide isomerase, among others. However, since the pioneering work of Anfinsen [4], it is known that a large number of water-soluble globular proteins are capable of reaching their native structure *in vitro* after being unfolded by changes in their environment, such as a raise of the temperature, the addition of denaturing agents or a change in the *pH*. It is the prediction of the native structure in these cases (only from the amino acid sequence and the laws of physics) that has become paradigmatic and receives the name of *protein folding problem*.

In 2005, in a special section of the Science magazine entitled "What don't we know?" [5], a selection of the hundred most interesting yet unanswered scientific questions was presented. What indicates the importance of the protein folding problem is not the inclusion of the question *Can we predict how proteins will fold?*, which was a must, but the large number of other questions which were related to or even dependent on it, such as *Why do humans have so few genes?*, *How much can human life span be extended?*, *What is the structure of water?*, *How does a single somatic cell become a whole plant?*, *How many proteins are there in humans?*, *How do proteins find their partners?*, *How do prion diseases work?*, *How will big pictures emerge from a sea of biological data?*, *How far can we push chemical self-assembly?* or *Is an effective HIV vaccine feasible?*, to quote just a few of them[1].

Some authors [6] divide the problem in two parts: the prediction of the three-dimensional, biologically functional, native state of a protein and the description of the actual folding process that takes the protein there from the unfolded state. The first part, which is more pressing and more technologically oriented, is included in the second part and it is, therefore, easier to tackle, as the relative success of knowledge-based methods suggests [7, 8]. However, we believe that, not only much theoretical insight may be gained from a solution of the more general second part of the problem, but also much engineering and design power, as well as new comprehension about so distinct topics as the ones quoted in the preceding paragraph. This is why our approach is one of bottom-top and ab initio flavor.

POTENTIAL ENERGY FUNCTIONS

The fundamental theory of matter that is nowadays accepted as correct by the scientific community is Quantum Mechanics. For the study of the conformational behaviour of a molecule consisting of n atoms, with atomic numbers Z_α and masses M_α, $\alpha = 1, \ldots, n$,

---

[1] We do not imply that the solution to the protein folding problem will solve any of these daunting scientific questions. Much collaboration and progress in many different disciplines is needed for that. However, the capability of predicting the native structure of proteins from the amino acid sequence may certainly constitute an important brick in these buildings.

one typically assumes that relativistic effects are negligible[2] and that, according to the Born-Oppenheimer scheme [9], the great differences in mass between electrons and nuclei allows to consider that the former are described by a Hamiltonian which is adiabatically decoupled from the nuclear one and that depends only parametrically on the positions of the nuclei. Hence, the behaviour of the system *in vacuum* may be extracted from the non-relativistic time-independent nuclear Schrödinger equation:

$$\left(-\sum_{\alpha=1}^{n} \frac{\hbar^2}{2M_\alpha} \nabla_\alpha^2 + \sum_{\beta>\alpha} \left(\frac{e^2}{4\pi\varepsilon_0} \right) \frac{Z_\alpha Z_\beta}{|\vec{R}_\beta - \vec{R}_\alpha|} + E_e^0(R) \right) \Psi_N(R) = E\,\Psi_N(R)\,, \quad (1)$$

where $E_e^0(\vec{R})$ denotes the effective potential due to the electronic cloud in the fundamental energy state[3] and R is shorthand for $\vec{R}_1,\ldots,\vec{R}_n$.

Despite the exponential growth in computing power that has been taking place in the last decades , a precise description of the behaviour of any biologically interesting system derived from the solution of (1) remains far from being even imaginable. Not to mention the huge complications that arise when the unavoidable inclusion of solvent is considered. This is why, omitting a myriad of possible intermediate descriptions, the most popular choice for the *in silico* prediction of the protein folding process has become the use of the so-called *force fields* [10–13], in which one assumes that the behaviour of the macromolecule (omitting again the solvent, to compare with (1)) is *classical* and may be described via a very simple potential energy function which, typically, has the form

$$V_{\text{ff}} := \frac{1}{2}\sum_{\alpha=1}^{N_r} K_{r_\alpha}(r_\alpha - r_\alpha^0)^2 + \frac{1}{2}\sum_{\alpha=1}^{N_\theta} K_{\theta_\alpha}(\theta_\alpha - \theta_\alpha^0)^2 + \sum_{\alpha=1}^{N_\phi} A_\alpha \cos(B_\alpha\phi_\alpha + \phi_\alpha^0) +$$

$$+ \sum_{\beta>\alpha} \left(\frac{C_{12}^{\alpha\beta}}{R_{\alpha\beta}^{12}} - \frac{C_6^{\alpha\beta}}{R_{\alpha\beta}^6} \right) + \sum_{\beta>\alpha} \left(\frac{e^2}{4\pi\varepsilon_0} \right) \frac{Z_\alpha Z_\beta}{R_{\alpha\beta}} \quad (2)$$

where r_α are bond lengths, θ_α are bond angles, ϕ_α are dihedral angles[4] and $R_{\alpha\beta}$ denotes the interatomic distances. Finally, all the parameters entering (2) (which may amount to thousands) are customarily fitted to reproduce thermodynamical measurements or taken from quantum mechanical calculations.

While it is true that these empirical potentials may be detailed enough to deal with simple conformational transitions in already folded proteins or with collective motions of systems of many proteins, and that they may also be used as scoring functions for

---

[2] Which, in organic molecules, is approximately correct for all the particles involved, except, maybe, for some core electrons in the heaviest atoms.
[3] This additional assumption that the electrons are in the fundamental state prevents us from describing the catalytic behaviour of most enzymes, however, the only interest here is to describe the folding process.
[4] For the sake of simplicity, no harmonic terms have been assumed for out-of-plane angles or for hard dihedrals, such as the peptide bond angle ω

protein design (as in the contribution by A. Jaramillo), all these applications require only that the energetics of the native structure and its surroundings be correctly described. As A. Tramontano explains in her contribution, the usefulness of these simple potentials for *de novo* structural prediction (assessed via the CASP contest[5]) remains much limited.

We believe that one of the reasons of this failure is the lack of accuracy of the potential energy functions used, since, even if the parameters fit is properly carried out, the choice of the very particular dependencies, for example those in (2), constitutes a heavy restriction in the space of functions. Accordingly, one of our aims is the design of classical potentials which are as similar as possible to the effective Born-Oppenheimer one in (1). To do this, one must calculate the electronic energy $E_e^0(R)$ using the powerful tools of Quantum Chemistry (see the contributions by J. J. Dannenberg and M. Amzel) and devise numerically efficient approximations to it.

In any case, in order to walk the long path connecting Quantum Mechanics and a classical description amenable to nowadays computers, one must have a meaningful way of comparing different approximations of the potential energy of a system. Much in the spirit of the contribution by M. Wall, and using the fact that the complex nature of biological molecules suggests the convenience of statistical analyses, we have designed in [14] a *distance*, denoted by d_{12}, between any two different potential energy functions, V_1 and V_2, that, from a working set of conformations, measures the typical error that one makes in the *energy differences* if V_2 is used instead of the more accurate V_1, admitting a linear rescaling and a shift in the energy reference.

This distance, which has energy units, presents better properties than other quantities customarily used to perform these comparisons, such as the energy RMSD, the average energy error, etc. It may be related to the Pearson's correlation coefficient by

$$d_{12} = \sqrt{2}\,\sigma_2(1 - r_{12}^2)^{1/2} \; . \tag{3}$$

Finally, due to its physical meaning, it has been argued in [14] that, if the distance between two different approximations of the energy of the same system is less than RT, one may safely substitute one by the other without altering the relevant dynamical or thermodynamical behaviour.

EFFECTS OF CONSTRAINTS

Another reason underlying the difficulties faced in the computational study of the protein folding problem is that the large number of degrees of freedom brings up the necessity to sample an astronomically large conformational space [15]. In addition, the typical timescales of the different movements are in a wide range and, therefore, demandingly small timesteps must be used in Molecular Dynamics simulations in order to properly account for the fastest modes [16], which lie in the femtosecond range; whereas the folding of a large protein may take seconds. In order to deal with these problems, one may naturally consider the reduction of the number of degrees of freedom describing macromolecules via the imposition of constraints.

---

[5] See http://predictioncenter.org

To manage this situation, we have made progresses in two directions. First, we have devised [17] a set of internal coordinates called *SASMIC* (standing for *Systematic and Approximately Separable and Modular Internal Coordinates*), which are much convenient to describe branched molecules and, specially, polypeptides, without having to rewrite the whole Z-matrix upon addition of new residues to the chain, and also allow to maximally separate the soft and hard movements[6].

Second, we have used these coordinates, the distance discussed before and the factorization of the external variables in the mass-metric determinants that we describe in [18], to study the possibility of neglecting the conformational dependence of the correcting terms that appear in the equilibrium distribution of organic molecules [19].

Constraining the hard coordinates q^I to be specific functions $f^I(q^i)$ of the soft ones (which defines a hypersurface Σ in the whole conformational space) produces two *classical* constrained models which are known to be conceptually [20, 21] and practically [22, 23] inequivalent: they are called *stiff* and *rigid*. In the classical rigid model, the constraints are assumed to be *exact* and all the velocities that are orthogonal to the hypersurface defined by them vanish. In the classical stiff model, on the other hand, the constraints are assumed to be *approximate* and they are implemented by a steep potential that drives the system to the constrained hypersurface. In this case, the orthogonal velocities are activated and may act as 'heat containers'.

The conformational equilibrium of the system, according to these models, is described by the following probability densities [19]:

Classical Stiff Model	Classical Rigid Model

$$P_s(q^u) = \frac{\exp\left[-\beta F_s(q^u)\right]}{Z'_s} \qquad\qquad P_r(q^u) = \frac{\exp\left[-\beta F_r(q^u)\right]}{Z'_r}$$

$$F_s(q^u) := V_\Sigma(q^i) - T\left(S_s^c(q^i) + S_s^k(q^u)\right) \qquad F_r(q^u) := V_\Sigma(q^i) - TS_r^k(q^u) \qquad (4)$$

$$S_s^k(q^u) := \frac{R}{2}\ln\left[\det G\left(q^u, f^I(q^i)\right)\right] \qquad S_r^k(q^u) := \frac{R}{2}\ln\left[\det g(q^u)\right]$$

$$S_s^c(q^i) := -\frac{R}{2}\ln\left[\det \mathcal{H}(q^i)\right]$$

where $\beta := 1/RT$, V_Σ is the potential energy in Σ (the Potential Energy Surface (PES) frequently used in Quantum Chemistry), and G, g and $\mathcal{H}$ denote, respectively, the whole-space mass-metric tensor, the reduced mass-metric tensor in Σ and the Hessian of the constraining part of the potential.

The different terms that correct the PES V_Σ in (4) are regarded (and denoted) as entropies because they are linear in the temperature T and come from the averaging out of certain degrees of freedom (sometimes coordinates, sometimes momenta). Accordingly, the effective potentials occurring in the exponent of the equilibrium probabilities are regarded (and denoted) as free energies.

---

[6] An automatic Perl script that generates the SASMIC Z-matrix, in the format of typical Quantum Chemistry packages, such as GAMESS or Gaussian03, from the sequence of amino acids, may be found at http://neptuno.unizar.es/files/public/gen_sasmic/

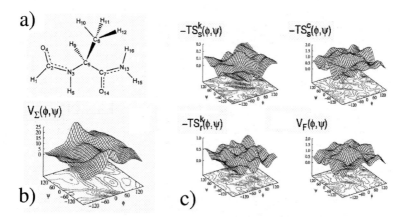

FIGURE 1. **a)** Model dipeptide HCO-L-Ala-NH$_2$ numbered according to the SASMIC [17] scheme. **b)** Potential Energy Surface. **c)** Conformational dependence of the correcting terms. All energies are given in kcal/mol.

Now, if Monte Carlo simulations in the coordinate space are to be performed [24, 25] and the probability densities that correspond to any of these two models sampled, the correcting entropies in (4) should be included or, otherwise, showed to be negligible.

On the other hand, if rigid Molecular Dynamics simulations are performed with the intention of sampling from the *stiff* equilibrium probability P_s [26–28], then, the so-called *Fixman's compensating potential* [29],

$$V_F(q^u) := TS_r^k(q^u) - TS_s^c(q^i) - TS_s^k(q^u) = \frac{RT}{2}\ln\left[\frac{\det G(q^u)}{\det \mathscr{H}(q^i)\det g(q^u)}\right], \quad (5)$$

must be added to the PES V_Σ.

The conformational dependence of most of the determinants appearing in (4) and (5) is frequently assumed to be negligible in the literature and they are consequently dropped from the calculations [30–33]. Also, subtly entangled to the assumptions underlying these simplifications, a second type of approximation is made that consists of assuming that the equilibrium values of the hard coordinates do not depend on the soft coordinates [31–34]. This has been argued to be only approximate even in the case of classical force fields [35–37].

In [19], we have eliminated all simplifying assumptions and measured the conformational dependence on the Ramachandran angles ϕ and ψ (the soft coordinates) of *all correcting terms* and of the Fixman's compensating potential in the model dipeptide HCO-L-Ala-NH$_2$. The potential energy function used was the effective Born-Oppenheimer potential for the nuclei (see (1)) derived from ab initio quantum mechanical calculations including electron correlation at the MP2/6-31++G(d,p) level of the theory.

In table 1, the main results of our work are presented. The importance of all the correcting terms is assessed by comparing (with the statistical distance d_{12} described in

TABLE 1. Quantitative assessment of the importance of the different correcting terms involved in the study of the constrained equilibrium of the protected dipeptide HCO-L-Ala-NH$_2$ (see [19]).

Corr.*	V_1†	V_2**	d_{12}‡	N_{res}§	b_{12}¶	r_{12}‖
$-TS_s^k - TS_s^c$	F_s	V_Σ	0.74 RT	1.82	0.98	0.9967
$-TS_s^c$	F_s	$V_\Sigma - TS_s^k$	0.74 RT	1.83	0.98	0.9967
$-TS_s^k$	F_s	$V_\Sigma - TS_s^c$	0.11 RT	80.45	1.00	0.9999
$-TS_r^k$	F_r	V_Σ	0.29 RT	11.62	1.01	0.9995
V_F	F_s	F_r	0.67 RT	2.24	0.97	0.9972

* Correcting term whose importance is measured in the corresponding row
† 'Correct' potential energy; the one containing the correcting term
** 'Approximate' potential energy; the one lacking the correcting term
‡ Statistical distance between V_1 and V_2 (see [14])
§ Number of residues in a polypeptide potential up to which the correcting term may be omitted
¶ Slope of the linear rescaling between V_1 and V_2
‖ Pearson's correlation coefficient

the previous section) the effective potential V_1, containing the term, with the approximate one V_2, lacking it. Moreover, if one assumes that the effective energies compared will be used to construct a polypeptide potential, the number N_{res} of residues up to which one may go keeping the distance between the two approximations of the the N-residue potential below RT is (see eq. (23) in [14]):

$$N_{res} = \left(\frac{RT}{d_{12}}\right)^2. \tag{6}$$

In the table, one can see that, in the stiff model, the Hessian-related correcting term should be included in Monte Carlo simulations for peptides as short as two residues, while the one that depends on G may be neglected up to chains which are ~ 80 residues long. The only correcting term occurring in the rigid model, in turn, may be dropped up to ~ 12 residues. Finally, the Fixman potential, containing all determinants, should be included in MD rigid simulations of peptides with more than two residues[7].

These results are related to a working set of conformations consisting of 144 points regularly distributed in the whole Ramanchandran space. In a second part of the work, we have repeated all the comparisons for a working set consisting of six secondary structure elements. The results suggest that, if one is interested only in this energetically lower region, the distances d_{12} are roughly divided by two and, accordingly, the values of N_{res} are four times larger.

---

[7] One should note that the distance between the PES V_Σ at MP2/6-31++G(d,p) and the one computed at HF/6-31++G(d,p) is $d_{12} \simeq 1.2\ RT$. A value slightly larger but of the order of the ones obtained when the most important correcting terms are dropped.

We have also repeated the calculations, with the same basis set (6-31++G(d,p)) and at the Hartree-Fock level of the theory in order to investigate if this less demanding method without electron correlation may be used in further studies. We have found that, indeed, this can be done, obtaining very similar results at a tenth of the computational effort.

As far as we are aware, this is *the first time* that this type of study is performed in a relevant biomolecule with a realistic potential energy function.

ACKNOWLEDGMENTS

We thank F. Falceto, V. Laliena, F. Plo and D. Zueco for illuminating discussions. The numerical calculations have been performed at the BIFI computing center. We thank I. Campos, for the invaluable CPU time and the efficiency at solving problems.

This work has been supported by the Aragón Government ("Biocomputación y Física de Sistemas Complejos" group) and by the research grants MEC (Spain) FIS2004-05073 and FPA2003-02948, and MCYT (Spain) BFM2003-08532. P. Echenique and I. Calvo are supported by MEC (Spain) FPU grants.

REFERENCES

1. C. M. Dobson, *Nature* **729**, 729 (2002).
2. J. W. Kelly, *Nat. Struct. Biol.* **9**, 323 (2002).
3. M. A. Basharov, *J. Cell. Mol. Med.* **7**, 223–237 (2003).
4. C. B. Anfinsen, *Science* **181**, 223–230 (1973).
5. Many authors, *Science* **309**, 78–102 (2005), http://www.sciencemag.com.
6. V. Daggett, and A. Fersht, *Nat. Rev. Mol. Cell Biol.* **4**, 497 (2003).
7. J. Moult, *Curr. Opin. Struct. Biol.* **15**, 285–289 (2005).
8. C. A. Rohl, C. E. Strauss, K. M. Misura, and D. Baker, *Methods Enzymol.* **383**, 66–93 (2004).
9. M. Born, and J. R. Oppenheimer, *Ann. Phys. Leipzig* **84**, 457–484 (1927).
10. B. R. Brooks, R. E. Bruccoleri, B. D. Olafson, D. J. States, S. Swaminathan, and M. Karplus, *J. Comp. Chem.* **4**, 187–217 (1983).
11. W. D. Cornell, P. Cieplak, C. I. Bayly, I. R. Gould, J. Merz, K. M., D. M. Ferguson, D. C. Spellmeyer, T. Fox, J. W. Caldwell, and P. A. Kollman, *J. Am. Chem. Soc.* **117**, 5179–5197 (1995).
12. W. L. Jorgensen, and J. Tirado-Rives, *J. Am. Chem. Soc.* **110**, 1657–1666 (1988).
13. T. A. Halgren, *J. Comp. Chem.* **17**, 490–519 (1996).
14. J. L. Alonso, and P. Echenique, *J. Comp. Chem.* **27**, 238–252 (2006).
15. K. A. Dill, *Prot. Sci.* **8**, 1166–1180 (1999).
16. T. Schlick, E. Barth, and M. Mandziuk, *Annu. Rev. Biophys. Biomol. Struct.* **26**, 181–222 (1997).
17. P. Echenique, and J. L. Alonso, *To appear in J. Comp. Chem.* (2006), (arXiv:q-bio.BM/ 0511004).
18. P. Echenique, and I. Calvo, *Submitted to J. Comp. Chem.* (2006), (arXiv:q-bio.QM/ 0512033).
19. P. Echenique, I. Calvo, and J. L. Alonso, *Submitted to J. Comp. Chem* (2006), (arXiv:q-bio.QM/ 0601042).
20. J. M. Rallison, *J. Fluid Mech.* **93**, 251–279 (1979).
21. E. Helfand, *J. Chem. Phys.* **71**, 5000 (1979).
22. D. Chandler, and B. J. Berne, *J. Chem. Phys.* **71**, 5386–5387 (1979).
23. M. Gottlieb, and R. B. Bird, *J. Chem. Phys.* **65**, 2467 (1976).
24. R. A. Abagyan, M. M. Totrov, and D. A. Kuznetsov, *J. Comp. Chem.* **15**, 488–506 (1994).
25. E. W. Knapp, and A. Irgens-Defregger, *J. Fluid Mech.* **14**, 19–29 (1993).
26. M. Pasquali, and D. C. Morse, *J. Chem. Phys.* **116**, 1834 (2002).
27. S. He, and H. A. Scheraga, *J. Chem. Phys.* **108**, 287 (1998).
28. M. Fixman, *J. Chem. Phys.* **69**, 1527 (1978).

29. M. Fixman, *Proc. Natl. Acad. Sci. USA* **71**, 3050–3053 (1974).
30. N. Gō, and H. A. Scheraga, *Macromolecules* **9**, 535 (1976).
31. A. Patriciu, G. S. Chirikjian, and R. V. Pappu, *J. Chem. Phys.* **121**, 12708–12720 (2004).
32. M. P. Allen, and D. J. Tildesley, *Computer simulation of liquids*, Clarendon Press, Oxford, 2005.
33. D. Frenkel, and S. B., *Understanding molecular simulations: From algorithms to applications*, Academic Press, Orlando FL, 2002, 2nd edn.
34. D. C. Morse, *Adv. Chem. Phys.* **128**, 65–189 (2004).
35. J. Zhou, S. Reich, and B. R. Brooks, *J. Chem. Phys.* **111**, 7919 (2000).
36. B. Hess, H. Saint-Martin, and H. J. C. Berendsen, *J. Chem. Phys.* **116**, 9602 (2002).
37. J. Chen, W. Im, and C. L. Brooks III, *J. Comp. Chem.* **26**, 1565–1578 (2005).

Multiscale Modeling of Tumor Cell Migration

Muhammad H. Zaman

Department of Biomedical Engineering, University of Texas at Austin
Austin, TX, 78712
mhzaman@mail.utexas.edu

Abstract. Most of our current knowledge of tumor cell migration comes from *in vitro* studies carried out on two-dimensional substrates, which provide only limited powers of observation. Moving to a 3D environment offers a superior range of observation, however the 3D experiments carried out to date have not studied the effect of matrix properties or cell-matrix interactions on cell migration. In order to fully understand the underlying mechanisms by which cells migrate *in vivo*, it is critical that we study the movement of cells in 3D environments that mimic the properties of the ECM *in vivo*. Using a combination of probabilistic, deterministic and atomistic modeling, we study the role of adhesion, mechanics, stiffness and matrix proteolysis on cell migration in 3D matrices. Our computational approaches combine novel techniques in ODE based modeling to calculate the forces acting on the cell during migration, Monte Carlo Simulations to study persistence and MD simulations to understand integrin-ECM interactions. Our techniques allow us to quantitatively map the cell migration landscape at various length and time scales. The results from our simulations show a simultaneous dependence of cell migration behavior on both chemical signals in the ECM and mechanical properties of the cell and the ECM. A complete understanding of this complex process requires that both the mechanics of the matrix and chemical signals be taken into account, as ignoring either of the two key components will lead to incomplete and inaccurate understanding of migration in 3D environments. The novel computational methods developed in our group show good agreement with the experimental studies in predicting the overall behavior of cells migrating in 3D and have provided useful insights in understanding the molecular basis of migration, contact guidance and persistence in 3D environments.

Keywords: Cell Migration; Molecular Dynamics; Cell Matrix Interactions; Extra Cellular Matrix.
PACS: 87.17.Jj ; 87.17.Aa

INTRODUCTION

Cell migration plays a key role in a wide variety of physiological processes. From early development to the maturity of an adult organism, and even afterwards, inability of the cells to migrate, or migration to incorrect locations leads to wide variety of disorders[1]. One such case, where cell motility plays a central role, is the invasion and spread of cancer. Most cancer related fatalities are due to tumor metastasis, a process during which cells move from the site of the tumor through the basement membrane to the vital organs via the blood stream. Malignant cancer cells are highly invasive and migrate through the surrounding extra cellular matrix (ECM) by adhering to the matrix proteins, generating traction and responding to intra and extra-cellular signals[2]. A thorough and quantitative understanding of tumor metastasis therefore requires systematic identification of the most prominent cellular and extra-cellular parameters that control cell motility in vivo. Most of the studies to date to study tumor cell motility have been predominantly qualitative and have focused on two-dimensional substrates. While these 2D theoretical and experimental efforts have provided us with useful information about biomechanics and biophysics of tumor cell motility, these studies are limited in their capacity to study the role of in vivo cell matrix adhesions,

CP851, *From Physics to Biology; BIFI 2006 II International Congress,*
edited by J. Clemente-Gallardo, Y. Moreno, J. F. Sáenz Lorenzo, and A. Velázquez-Campoy
© 2006 American Institute of Physics 0-7354-0350-3/06/$23.00

are blind to matrix proteolysis and introduce artificial polarities in cells. A description of invasion based on artificial substrates is going to paint an incomplete picture at best, and will most likely result in a misleading and an inaccurate understanding. Only recently, a few groups have utilized high-resolution imaging and spectroscopic techniques to quantitatively study cell migration in native like environments [3-7]. In addition, a few efforts have also been made to study various aspects of cell-cell and cell-matrix interactions in native like environments[3, 8, 9]. These studies, combined with high- resolution experimental methods, show great promise in developing a thorough understanding of cancer at multiple length and timescales and will ultimately aid in design of more effective therapeutics.

The present study focuses on development and implementation of multiscale computational approaches to study tumor cell migration. The basic theme is to develop and apply computational methods rooted in classical and statistical mechanics to understand tumor cell motility at multiple length and timescales. While there are several limitations of the current multiscale approach, the ability of these approaches to make quantitative predictions and provide a first principle understanding of tumor cell-matrix interactions is highly desirable to develop a thorough understanding of underlying biophysical processes governing tumor cell motility.

METHODS

The details of the methods employed in our multiscale modeling have been described in detail elsewhere[8, 9]. In this section, we only briefly discuss the methods utilized to study cell-migration in native like three-dimensional environments. The current study focuses on only two broad aspects of our multiscale model, namely cell-matrix interactions at a single molecule level, mean-field models of single and multiple cells in 3D environments.

Cell-Matrix Interactions at the Molecular Level: To understand the role of conformational changes in the extracellular matrix, and its effect on cell adhesion and migration, we carry out multiple long-time (>100ns) MD and SMD simulations on integrin-collagen complex [10] using Charmm and OPLS force fields. These simulations mimic the conformational changes in ECM occurring routinely due to cell migration, proteolysis and adhesion. Small conformational changes are introduced in the simple collagen-like protein by mechanical stretching and the binding of stretched and unstretched collagen to integrin is studied using free energy perturbation methods. Using molecular dynamics methods, we also study the interactions between integrin and denatured collagen, which serves as a reasonable model for integrin-gelatin (denatured collagen) interactions. In addition we also study the dynamics of key amino acids at the binding interface and study the role of mutations both in the collagen peptide and integrin binding surface on integrin-collagen interactions.

Figure1. Collagen-Integrin interface from x-ray structure described in ref. 10 The oval shows the binding interface region and the arrows point to the direction of collagen stretching in the simulation.

Cellular Models: Mean-Field single cell models are constructed to study the role of adhesion, matrix sterics and mechanical properties of the ECM. The details of the model are outlined elsewhere [8, 9]. The model is designed to predict cell migration as a function of time by calculating the forces acting on the centroid at each time step (Δt). The time step (Δt) in our simulation is equal to the time taken by the leading edge to produce stable protrusions, adhere to the ligands in the matrix and move in the direction of protrusion. The total force is broadly divided into traction (rear and front), leading edge protrusion and viscous drag in 3D. The traction force (at front or back) can be mathematically represented as:

$$\mathbf{F}_{trac} = F_{R-L} \times \beta \quad (1)$$

Thus the traction force at the front and the rear depend upon the force per ligand-receptor complex (F_{R-L}), which is a function of the Young's modulus, E_{mod}, of the surrounding medium. To account for the possibility of different numbers of receptors at the front and the back of the cell, and/or difference in binding strength of these receptors to the ligand, we introduce β, a dimensionless parameter measuring the binding strength of the receptors to the ligands in the ECM. β is defined as:

$$\beta = k_1 \times n \times [L] \quad (2)$$

where n is total number of available receptors, [L] is the concentration of the ligands in the ECM (in the units of M), and k is the binding constant for the binding of integrins ligands in the ECM (units of M^{-1}).

For the front and rear end we can write:

$$\beta_f = k_1 \times n_f \times [L_f] \quad (2a)$$
$$\beta_b = k_1 \times n_b \times [L_b] \quad (2b)$$

For our model we assume that the as the cell polarizes, integrins are distributed asymmetrically on the cell surface, i.e. $n_f > n_b$. As a first approximation, we can ignore proteolysis effects (if the pore size of the matrix is relatively large), however we can incorporate proteolytic activity at the leading edge that would result in $L_f \neq L_b$.

The second force, $\mathbf{F}_{drag}$ arising from viscous resistance to movement. Proportional to the cell movement speed, the drag force is given by:

$$\mathbf{F}_{drag} = c \times \eta v \quad (3)$$

where η is the effective viscosity of the viscoelastic medium (considered a constant throughout the matrix) and v is the velocity. The constant "c" depends on the shape of cell, for a spherical cell in an infinitely viscous medium, $c = 6\pi \times$ cell radius.

$\mathbf{F}_{protrusion}$ is generated by actin polymerization at the leading edge and the order of magnitude estimate is determined from previous experimental studies[11, 12]. The direction of $\mathbf{F}_{protrusion}$ is chosen randomly at each time step. The protrusion force in our model is in fact, a time averaged value of positive protrusive forces over Δt.

The total force acting on the cell is therefore given by:

$$\mathbf{F}_{tot} = \mathbf{F}_{drag} + \mathbf{F}_{trac} + \mathbf{F}_{protrusion} = 0 \quad (4)$$

Our model uses Equation 4 to calculate the cell velocity within its movement environment for each simulation cycle, or time-step. In each cycle, the cell protrudes in a particular direction with F_{prot}, makes attachments based upon the k_1, n_f and n_b and experiences resistive forces due to viscoelastic resistance.

RESULTS:

Integrin binding to native and denatured collagen. Using free energy perturbation methods employing long-time molecular dynamics simulations (> 100 ns), we study the role of conformational changes in collagen on the overall binding between collagen and integrin. The results of our simulations suggest that collagen-integrin binding is adversely affected as collagen is mechanically stretched and results in a net loss of stability of 36 ± 5.5 kcal of stability as the protein is unfolded. This result is consistent within the error margin with experimental observations [10]. To get a molecular insight into the results, we studied the role of amino acids at the binding interface, namely, D151, S153, T221, D219, Q215 and D254. We observe that these amino acids, which have been shown to be critical for stability, are unable to form hydrogen bonds with denatured collagen due to conformational changes in collagen. In addition, structural changes in collagen force these amino acids to predominantly occupy extended conformation whereas in the bound state these amino acids prefer helical conformations.

Dependence of cell speed on ligand density and adhesivity. The single and multi-cellular simulations are more coarse-grained than the atomistic simulations described briefly above. We utilize data from experimental and computational methods (such as the ones described above) and incorporate them into our equations describing ligand-receptor binding in three dimensional matrices. Our results suggest the presence of a bimodal behavior of cell speed three dimensional matrices on receptor density and ligand concentration (Figure 3). The bimodal behavior occurs due to a balance between adhesion, ligand concentration and maximum speed. At low ligand density, there is not enough adhesion for the cells to generate traction, while at high ligand concentration the cells are able to attach but do not detach and hence do not move greater distance. In addition, we notice that as the receptor-ligand density is weakened the bimodal behavior is shifted down. Thus in a molecular model system studied above, where molecular interactions are weakened by either mutations or collagen stretching, maximum speed is affected and is lowered. These results are important as mutations and conformational changes occur often in physiological systems and affect the ability of cells to adhere and migrate.

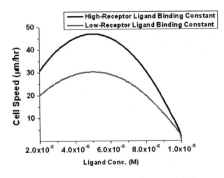

Figure 2. Cell speed shows a bimodal behavior with change in ligand density and reaches a maximum at intermediate ligand density. Decrease in integrin-collagen binding results in decrease in maximum speed.

Multidimensional speed landscapes. A key aspect of our simulations is observing the simultaneous dependence of cell speed and persistence on multiple parameters. To capture the complexity of in vivo situations, where multiple parameters vary simultaneously, we also study the role of multiple parameters varying simultaneously. One such situation is depicted in Fig 3 where cell speed surface is shown as a function of matrix stiffness and ligand density. The figure shows that individual slices of this surface can be highly misleading as the cell speed may have very different behaviors in response to small variations in cellular or extracellular parameters. Landscape figures such as these underline the importance of system-wide understanding and analysis of cellular systems where multiple parameters are being affected simultaneously.

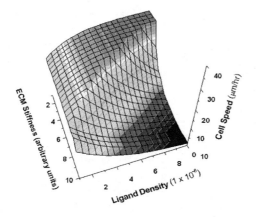

Figure 3. Multidimensional landscape of cell speed plotted simultaneously with ligand density and matrix stiffness.

The results of our simulations show very good agreement with experimental results [3-7]. Our simulations not only predict the order of magnitude loss in binding between stretched collagen and integrin, they also provide a previously unknown molecular basis for the loss in stability. Our predictions of bimodal behavior in cell speed as a function of ligand concentration and adhesion have also been seen in novel 3D experiments [3, 4].

Putting it all together: Conclusions and Future Outlook.

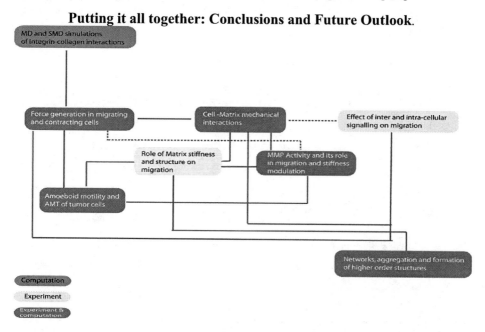

Figure 4. Integrated computational and experimental scheme employed in our group to study tumor cell migration and cell-matrix interactions at multiple length and timescales.

121

The present paper combines the results of computational methods at varying length and timescales to understand tumor cell motility in native and native like matrices. These methods rooted in statistical and classical physics have been used successfully in a number of biologically relevant problems. While any multiscale approach to understand complex problems such as cancer cell motility will many limitations and numerous approximations (which are discussed in details in references 8 and 9), nonetheless, we believe a system-wide multiscale approach, such as ours, has great potential in predicting the systems biology of cell migration. By switching from an atomistic to continuum approach depending on the length and timescale we are able to develop a consistent and unified framework that is able to address a number of highly relevant problems in tumor cell migration. Finally, our simulations work hand in hand with experimental studies to not only validate the predicted results but also to suggest new experimental tools aimed at developing a thorough understanding of tumor cell migration at multiple length and timescales. We hope that our multiscale approach, combined with state of the art interdisciplinary experiments will benefit researchers in developing more efficient therapeutics aimed at controlling and curing cancer.

ACKNOWLEDGMENTS

The author is grateful to Prof. D. Lauffenburger and Prof. P. Matsudaira for numerous enlightening discussions. MHZ gratefully acknowledges the support of Sokol Foundation Fellowship for Cancer Research.

REFERENCES

1. Webb, D.J. and Horwitz, A.F., *New dimensions in cell migration.* Nat Cell Biol, 2003. **5**(8): p. 690-2.
2. Lauffenburger, D.A. and Horwitz, A.F., *Cell migration: a physically integrated molecular process.* Cell, 1996. **84**(3): p. 359-69.
3. Shreiber, D.I., Barocas, V.H. and Tranquillo, R.T., *Temporal Variations in Cell Migration and Traction during Fibroblast-Mediated Gel Compaction.* Biophys J, 2003. **84**(6): p. 4102-14.
4. Lutolf, M.P., Lauer-Fields, J.L., Schmoekel, H.G., Metters, A.T., Weber, F.E., Fields, G.B. and Hubbell, J.A., *Synthetic matrix metalloproteinase-sensitive hydrogels for the conduction of tissue regeneration: engineering cell-invasion characteristics.* Proc Natl Acad Sci U S A, 2003. **100**(9): p. 5413-8.
5. Raeber, G.P., Lutolf, M.P. and Hubbell, J.A., *Molecularly engineered PEG hydrogels: a novel model system for proteolytically mediated cell migration.* Biophys J, 2005. **89**(2): p. 1374-88.
6. Wolf, K., Mazo, I., Leung, H., Engelke, K., von Andrian, U.H., Deryugina, E.I., Strongin, A.Y., Brocker, E.B. and Friedl, P., *Compensation mechanism in tumor cell migration: mesenchymal-amoeboid transition after blocking of pericellular proteolysis.* J Cell Biol, 2003. **160**(2): p. 267-77.
7. Friedl, P. and Brocker, E.B., *The biology of cell locomotion within three-dimensional extracellular matrix.* Cell Mol Life Sci, 2000. **57**(1): p. 41-64.
8. Zaman, M.H., Trapani, L.M., Sieminski, A.L., Wells, A., Kamm, R.D., Lauffenburger, D.A. and Matsudaira, P., *Cell Migration in Three-dimensional Matrices Is Inversely-dependent on Cell-matrix Adhesiveness and Matrix Stiffness.* Submitted, 2005.
9. Zaman, M.H., Kamm, R.D., Matsudaira, P. and Lauffenburger, D.A., *Computational model for cell migration in three-dimensional matrices.* Biophys J, 2005. **89**(2): p. 1389-97.
10. Emsley, J., Knight, C.G., Farndale, R.W., Barnes, M.J. and Liddington, R.C., *Structural basis of collagen recognition by integrin alpha2beta1.* Cell, 2000. **101**(1): p. 47-56.
11. Harris, A.K., Wild, P. and Stopak, D., *Silicone rubber substrata: a new wrinkle in the study of cell locomotion.* Science, 1980. **208**(4440): p. 177-9.
12. James, D.W. and Taylor, J.F., *The stress developed by sheets of chick fibroblasts in vitro.* Exp Cell Res, 1969. **54**(1): p. 107-10.

Variable-Barrier Modeling of Equilibrium Protein Folding

Jose M. Sanchez-Ruiz

Facultad de Ciencias, Departamento de Quimica Fisica, Universidad de Granada. Fuentenueva s/n, 18071-Granada, Spain and Biocomputing and Physics of Complex Systems research Institute, Corona de Aragon 42, Edificio Cervantes, E-50009 Zaragoza, Spain

Abstract. Deviations from two-state behavior in the equilibrium folding/unfolding of small proteins are likely due to small or vanishing thermodynamic free-energy barriers, rather than to the presence of significantly populated intermediate macrostates. A phenomenological variable-barrier model for equilibrium protein folding is described. Analysis of experimental data on the basis of this model leads to kinetically-relevant free-energy barriers and allows potential cases of global downhill folding to be identified.

Keywords: Protein folding equilibrium, protein folding kinetics, free-energy barriers, downhill folding.
PACS: 82.35.Pq 87.15Cc 87.15.He

THERMODYNAMIC BARRIERS AND DEVIATIONS FROM TWO-STATE BEHAVIOR

Experimental equilibrium data on protein folding/unfolding are usually interpreted in terms of well-defined macrostates. The paradigmatic example of this approach is the two-state model, in which only the native and unfolded macrostates (N and U) are assumed to be significantly populated and the folding/unfolded process is modeled as a simple chemical equilibrium:

$$N \leftrightarrow U \tag{1}$$

Deviations from two-state behavior (i.e., when a good fit to the experimental equilibrium profiles cannot be achieved on the basis of the model) are often put down to the existence of additional, significantly populated protein macrostates (equilibrium intermediate states). This interpretation certainly seems plausible in the case of large, complex proteins, with structures possibly made up of several, quasi-independent domains, since intermediate states in which certain domains are folded and other domains are unfolded will likely occur. On the other hand, for reasons we expound below, the interpretation of deviations from two-state behavior in terms of intermediate macrostates appears far less likely in the case of small proteins (in particular, in the case of small, *fast-folding* proteins).

Protein folding-unfolding involves the formation/breaking of a large number of weak, non-covalent interactions. Therefore, it is possible to conceive protein conformations (microstates) with essentially any degree of unfolding (from 0% to

CP851, *From Physics to Biology; BIFI 2006 II International Congress*,
edited by J. Clemente-Gallardo, Y. Moreno, J. F. Sáenz Lorenzo, and A. Velázquez-Campoy
© 2006 American Institute of Physics 0-7354-0350-3/06/$23.00

100%). Clearly, for two-state behavior to be observed, comparatively high free energies must correspond to degrees of unfolding intermediate between those of native and unfolded and protein. Thus, since the population of a given microstate is proportional to a Boltzmann exponential of its free energy, intermediate degrees of unfolding will not be significantly populated and two distinct populations (corresponding to the native and unfolded macrostates) will be observed (see Figure 1). I refer to the higher free energy assumed for intermediate degrees of unfolding as a *thermodynamic free energy barrier* (for lack of a better name). The main consequence of a high thermodynamic barrier is to produce a bimodal distribution with two well-defined macrostates (Figure 1). Whether such thermodynamic barrier has kinetic consequences for the rates of folding/unfolding is an entirely different issue which will be discussed further below.

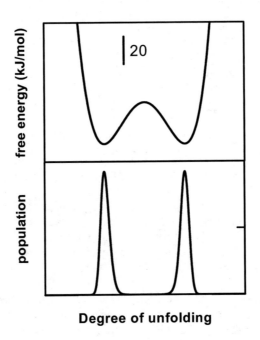

Degree of unfolding

FIGURE 1. A free energy profile with a significant barrier (upper panel) produces a bimodal distribution in a plot population of microstates versus degree of unfolding (lower panel). Such bimodal distribution is consistent with a two-macrostates scenario.

The picture described above could be easily extended to include additional intermediate macrostates, but this would necessarily mean including additional thermodynamic free energy barriers. For instance, for a three-macrostates equilibrium model,

$$N \leftrightarrow I \leftrightarrow U \qquad (2)$$

two thermodynamic barriers are required in order to "separate" N from I and I from U.

In the case of small proteins, however, the existence of several thermodynamic barriers does not seem likely, since it would imply large increases and decreases in free energy upon moderate changes in degree of unfolding. Actually, deviations from two-state behavior (eq. 1) with small proteins are more likely due to a low or vanishing thermodynamic barrier. Note that when the height of the thermodynamic barrier is on the order of the thermal energy (around 2.5 kJ/mol) all degrees of unfolding are significantly populated. In fact, the barrier could be non-existing altogether, which would imply a unimodal distribution in a plot of free energy versus degree of unfolding (figure 2). This case corresponds to the global downhill folding regime, for which experimental evidence has been recently reported (1-3).

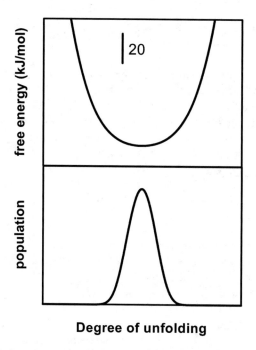

Degree of unfolding

FIGURE 2. A free energy profile with a negligible or non-existing barrier (upper panel) produces an unimodal distribution in a plot population of microstates versus degree of unfolding (lower panel). Such unimodal distribution is consistent with a global downhill scenario.

VARIABLE-BARRIER MODELING OF EQUILIBRIUM FOLDING/UNFOLDING

From the reasoning expounded above, it appears plausible that deviations from two state behavior occur often in small proteins, but not due the presence of additional intermediate macrostates, but to a marginal or vanishing thermodynamic barrier and the consequent approaching of the process to the single-macrostate, global-downhill folding regime. Certainly, the "chemical-like" models currently used in the analysis of equilibrium folding/unfolding data (such as those depicted in equations 1 and 2) do not take this possibility into account. However, modeling of protein folding/unfolding including a variable thermodynamic barrier is possible on the basis of physical models inspired by Landau phenomenological theory of critical transitions. For instance, the gas and liquid phases of a given substance can coexist at equilibrium (as two distinct phases) for temperatures and pressures in the liquid-vapor equilibrium line. As temperature and pressure are increased along that line, liquid and gas become more similar and eventually merge into a single phase at the critical point (for a pictorial illustration, see ref. 4). In the classical Landau theory of critical transitions (see chapter 10 in ref. 5) this is phenomenologically described with a free energy functional expressed as a series expansion in powers of an "order parameter" (the thermodynamic quantity that exhibits large fluctuations near critical conditions) and truncating the expansion at the quartic level. The truncated expansion produces a free energy functional with one or two free energy minima depending on the sign of the coefficient of the quadratic term.

A Landau free energy functional can be implemented in experimental differential scanning calorimetry (DSC) data analysis (6) by i) using a suitably defined enthalpy scale as an "order parameter", ii) writing the probability density for enthalpy microstate occupation at a given characteristic temperature (T_0) in terms of the enthalpy-dependence of the free energy,

$$P(H) = C \cdot \exp\left(- G_0(H) / RT_0\right) \tag{3}$$

and iii) expressing the free energy as,

$$G_0(H) = -2\beta \left(\frac{H}{\alpha}\right)^2 + |\beta| \left(\frac{H}{\alpha}\right)^4 \tag{4}$$

which is actually the Landau functional with the coefficients of the H^2 and H^4 expressed in terms of two parameters, α and β, which have a clear and intuitive meaning. Thus, it can be easily shown (see ref. 2 for details) that, for $\beta > 0$, $G_0(H)$ has a maximum at $H=0$ and two minima at $H=\pm\alpha$. Therefore, for $\beta > 0$, there are two macrostates with an enthalpy difference of about 2α and, in this case, β corresponds to the height of the free energy barrier separating the two mimima at the characteristic temperature. For $\beta \leq 0$, the free energy profile shows only a minimum and there is only one macrostate. In this case, α and β are just convenient parameters that describe the shape of the free energy functional. Therefore, it is just the *sign* of the parameter β which determines the observation of two macrostates or a single macrostate at the characteristic temperature T_0. Of course, positive but very small values of β (significantly smaller than the thermal energy, RT) are essentially equivalent to the single-macrostate, barrierless case. A final modification is introduced in the free energy functional to take into account the asymmetry expected in a folding-unfolding process. Such modification is explained in some detail in ref. 2.

The important point, however, is that the above approach allows us to calculate heat capacity versus temperature profiles for different values of the thermodynamic folding/unfolding barrier (β) following standard procedures (2,6). The implication of this calculation is the possibility of setting a non-linear, least-squares procedure to fit experimental DSC profiles for protein unfolding in which the height of the folding barrier is a fitting parameter. Such procedure does not impose the existence of two well-defined macrostates but, instead, the two-macrostate or single-macrostate character of the thermal unfolding process is an outcome of the analysis. In fact, in the first application of this *variable-barrier* phenomenological model (2), a significant barrier was determined for the two-state protein *E. coli* thioredoxin. On the other hand, essentially no barrier was found for the small, fast-folding protein BBL, which had been previously characterized as a downhill folder (1).

KINETIC IMPLICATIONS OF THERMODYNAMIC FOLDING BARRIERS

In principle, the barriers for folding/unfolding derived from variable-barrier analysis of DSC thermograms would be expected to have only thermodynamic consequences. That is, the existence of a large thermodynamic barrier (significantly larger than the thermal energy) guarantees that the population of conformations of intermediate degree of unfolding is very low, in such a way that two distinct macrostates are observed. Conversely, a low or non-existent thermodynamic barrier implies a unimodal distribution in a plot of population vs. degree of unfolding and, consequently, a single macrostate and global downhill folding behavior.

It may not be clear a priori whether such thermodynamic barriers have kinetic implications or not. Thus protein folding involves myriads of conformational coordinates and it is not clear that folding kinetics can be described in terms of a single reaction coordinate and, furthermore, that the enthalpy scale used in variable-barrier analysis is a good approximation to that hypothetical single reaction coordinate. It is therefore rather surprising and highly significant that an excellent correlation between folding rates and thermodynamic barriers derived from variable-barrier analysis of DSC data has been recently found (3). The study was based in a set of 15 proteins of similar size (64±15 residues) including examples of the three main structural classes and spanning about four orders of magnitude in folding times. The correlation coefficient was $r^2=0.9$ and the fastest folding proteins were found have essentially negligible (or non-existing) thermodynamic folding barriers, thus supporting the downhill character of their folding. Overall, the correlation suggests the exciting possibility of determining absolute, kinetically-relevant free-energy barriers from the analysis of equilibrium folding/unfolding data.

ACKNOWLEDGMENTS

Research in the author's lab is supported by Grant BIO2003-02229 from the Spanish Ministry of Education and Science, FEDER funds and Grant CVI-771 from the "Junta de Andalucia".

REFERENCES

1. M. M. Garcia-Mira, M. Sadqi, N. Fisher, J. M. Sanchez-Ruiz and V. Muñoz, *Science* **298**, 2191-2195 (2002).
2. V. Muñoz and J. M. Sanchez-Ruiz, *Proc. Natl. Acad. Sci. USA* **101**, 17646-17651 (2004).

3. A. N. Naganathan, J. M. Sanchez-Ruiz and V. Muñoz, *J. Am. Chem. Soc.* **127**, 17970-17971 (2005).
4. J. V. Senger, and A. L. Senger, *Chem. Eng. News* **46**, 104-118 (1968).
5. H. B. Callen, *Thermodynamics and an Introduction to Thermostatistics*, Wiley, New York, 1985.
6. B. Ibarra-Molero and J. M. Sanchez-Ruiz, "Differential scanning calorimetry of proteins: an overview and some recent advances" in *Advanced Techniques in Biophysics,* edited by J. L. Arrondo and A. Alonso, Elsevier, 2006, pp. 27-48.

FurA from *Anabaena* PCC 7120: New insights on its regulation and the interaction with DNA

J.A. Hernández [1], S. López-Gomollón [1], S. Pellicer [1], B. Martín [1], E. Sevilla [1], MT. Bes [1,2], M.L. Peleato [1,2] and M.F. Fillat [1,2].

(1) Departamento de Bioquímica y Biología Molecular y Celular. Facultad de Ciencias. (2) Instituto de Biocomputación y Física de Sistemas Complejos. Universidad de Zaragoza. 50009-Zaragoza. España

Abstract. Fur (ferric uptake regulator) proteins are global regulatory proteins involved in the maintenance of iron homeostasis. They recognize specific DNA sequences denoted iron boxes. It is assumed that Fur proteins act as classical repressors. Under iron-rich conditions, Fur dimers complexed with ferrous ions bind to iron boxes, preventing transcription. In addition to iron homeostasis, Fur proteins control the concerted response to oxidative and acidic stresses in heterotrophic prokaryotes. Our group studies the interaction between Fur proteins and target DNA sequences. Moreover, the regulation of FurA in the nitrogen-fixing cyanobacterium *Anabaena* sp. PCC 7120, whose genome codes for three *fur* homologues has been investigated. We present an overview about the different factors involved in the regulation of FurA and analyze the parameters that influence FurA-DNA interaction in the cyanobacterium *Anabaena* PCC 7120.

Keywords: iron, cyanobacteria, transcriptional regulation, Fur, protein-DNA interaction

INTRODUCTION

Iron is an essential nutrient for microorganisms survival since it takes part in many enzymes involved in different metabolic pathways (1). It is stated that iron deficiency in aquatic ecosystems might determine the ocean primary productivity, limiting both cell division and the quality and abundance of phytoplancton (2). In the case of cyanobacteria, the request for this element is particularly elevated due to the high number of iron-dependent metalloproteins present in the photosynthetic electron transport chains as well as in the nitrogenase complex in nitrogen-fixing species. Moreover, iron limitation induces the expression of iron-chelating siderophores, a chlorophyll-binding protein, and causes the replacement of proteins containing iron cofactors for non-iron proteins, such as flavodoxin (3). Other processes, such as the red tides or the cyanobacterial blooms, as well as cyanotoxin production have also been suggested to be related to the availability of iron (4). On the other hand, iron catalyses free radical formation in the presence of oxygen that could damage DNA, proteins and membrane lipids. Consequently, iron homeostasis must be kept in a strict equilibrium that needs an accurate control of the uptake, metabolism and storage of this element. In prokaryotic organisms, the Fur (ferric uptake regulation) protein

CP851, *From Physics to Biology; BIFI 2006 II International Congress,*
edited by J. Clemente-Gallardo, Y. Moreno, J. F. Sáenz Lorenzo, and A. Velázquez-Campoy
© 2006 American Institute of Physics 0-7354-0350-3/06/$23.00

family is an essential node that integrate iron metabolism and modulation of the free radical removal systems (1). At the moment, around 600 homologues have been described in different bacterial species (http://pfam.wustl.edu/cgi-bin/getdesc?acc=PF01475). The genome of the cyanobacterium *Anabaena* sp. PCC 7120 codes for 3 Fur-homologues that have been called FurA, FurB and FurC (5).

It is well described that in most cases, the DNA-binding activity of Fur is dependent on iron, acting Fe^{+2} as corepressor. Under iron-rich conditions, Fur is complexed with Fe^{+2} and binds to the iron-boxes located in the promoter regions of many iron-regulated genes, thus preventing transcription. Under iron starvation, Fur loses iron and dissociates of DNA, allowing transcription to occur. In *E. coli*, most iron-boxes show a 19 bp palindromic sequence GATAATGATAATCATTATC which has been re-evaluted as, at least, three adjacent NAT(A/T)AT repeats which can be extended by addition of further hexameric units in any orientation (6). This A,T-rich consensus sequence seems to be highly conserved, showing a higher degree of identity in the genera phylogenetically closer. Fur requirements for binding can be different depending of the organism. The need of the presence of Fe^{+2} or Mn^{+2} for Fur activity is the most frequent situation. However, when optimal conditions for Fur-DNA interaction from the cyanobacterium *Anabaena sp.* were explored, absolute metal or redox state requirements were not found (7). However, presence of divalent metals and/or reducing conditions provided by 1mM DTT improved the affinity of Fur for their putative consensus sequences. Whether the protected sequences of the operator are identical with or without metals and depending on the reducing conditions is a very interesting topic that remains to be investigated. Moreover, oligomerization degree can be a key factor in the ability of the protein to bind DNA (8).

Expression of Fur proteins is regulated by different factors depending on the microorganism. The most common type of Fur control is a moderate auto-regulation. The modulation of Fur proteins seems to be particularly complex in *Anabaena* PCC 7120 where we have described that there is cross-regulation between the Fur family members (5). Another important factor that modulates the FurA concentration inside the cell is the α-*furA* antisense RNA (9). In the same way, heme presence induces a substantial decrease in the FurA affinity for its target sequences (10) (Figure 1).

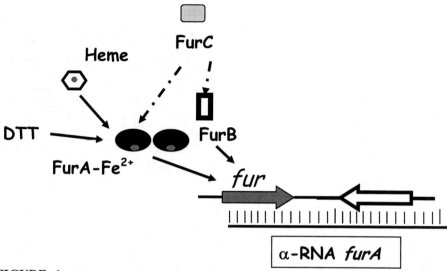

FIGURE 1. Scheme showing regulation of FurA that takes place at transcriptional, post-transcriptional and post-translational levels.

In this work, we have determined the affinity of FurA for its target DNA sequences PfurA and PisiB and a possible regulation by phosphorylation of native FurA in *Anabaena*, has been investigated.

EXPERIMENTAL PROCEDURES

Overexpression and partial purification of FurA were performed as described in (7). Native FurA from *Anabaena* and the recombinant protein from *E. coli* extracts were purified by chromatography through a heparin-Sepharose 6 Fast Flow column (Amersham-Pharmacia) according to (8). Electrophoretic mobility shift assays were performed as described in (7) using the 389 bp and 348 bp fragments from the upstream regions of furA and isiB genes respectively amplified by PCR. Results were processed with a Gel Doc 2000 image analyser from BioRad. For calculations of Kd(app), estimation of remaining unbound DNA in each sample was calculated with respect to the band area corresponding to free DNA control, taken as 100%.

RESULTS AND DISCUSSION

Interaction Of FurA-DNA In *Anabaena*: Binding To *furA* And Flavodoxin Promoters.

Promoter regions for *furA* and flavodoxin (*isiB*) promoters have been defined previously (7). We have investigated the affinity of FurA for both promoters using EMSA assays of FurA versus each promoter region in the presence of an unspecific competitor DNA (7) and performing competition assays between both promoters.

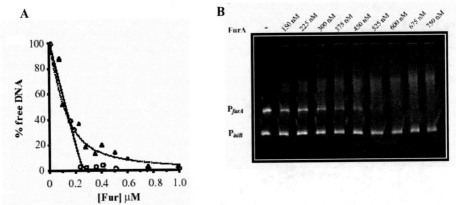

FIGURE 2. A.- Binding of FurA to *furA* and *isiB* promoters. A. Determination of $K_{d(app)}$ for the binding of FurA to P_{furA} (circles) and P_{isiB} (triangles). P_{furA} 12 μM was titrated with FurA (100, 150, 175, 225, 275, 350, 400, 500, 750 and 1000 nM) as described previously using the core buffer supplemented with 100 μM $MnCl_2$ and 1 mM DTT. (100, 150, 175, 225, 275, 350, 400, 500, 750 and 1000 nM) P_{isiB} was titrated in the same conditions with FurA 100, 150, 225, 275, 350, 400, 500, 600, 750 and 1000 nM. **B.** Competition of the *Anabaena fur* and *isiB* promoter fragments for FurA binding. Electrophoretic mobility shift assay of equimolar amounts of $P_{furA \ and} P_{isiB}$ in the presence of increasing amounts of FurA. Binding was performed in the presence of 100 μM $MnCl_2$ and 1 mM DTT.

Figure 2A shows a comparative analysis of the titration curves of P_{furA} and P_{isiB} with FurA. Apparent Kd (Kd(app)), is defined as the Fur concentration at which 50% of the DNA is associated with the protein. The validity of those equilibrium-binding constants depends critically on the proportion of active protein in the FurA preparation, which can change mainly due to the tendency of the repressor to oligomerize (8). Therefore, the titration assays shown in figure2 have been been carried out a minimum of 3 times, leading to repetitive results. Optimal circumstances for Fur-DNA binding were observed when Mn(II) and DTT were added in the assays. Under those conditions, data were fitted to rectangular hyperbolas to estimate the apparent repressor-operator dissociation constants [Kd(app)] for PfurA and PisiB (Fig. 2). Resulting values were considerably higher for PfurA, whose [Kd(app)] was 0.49 + 1.17 nM versus [Kd(app)]= 52.79 + 13.4 nM obtained for the FurA-PisiB interaction. Figure 2B shows competitive binding assays incubating increasing amounts of FurA protein with an equimolar mixture of furA and isiB promoters. As it was expected those results confirm that, in vitro, FurA binds its own promoter with higher affinity than PisiB.

Study Of The Oligomerization Of FurA.

There are several reports stating the importance of the oligomerization state of DNA-binding proteins in their affinity for target DNA sequences (8,11). We have investigated the state of the oligomerization of FurA using purified recombinant protein. Native PAGE of FurA was not possible to be performed by the usual procedures, due to precipitation of the protein at neutral-basic pHs. However, an alternative method at acidic pHs was successfully used, and allowed us to observe the presence of several bands (Figure 3) with different mobility. The different bands correspond, presumably, to the presence of oligomeric forms, and the oligomerization pattern is very different in two samples from different Fur concentration. As shown in

Figure 3, sample in lane 1 came from a dilute Fur stock (70 μM), while sample in lane 2 corresponds to a concentrate Fur solution (600 μM). The two bands in lane 1 could be coincident with the to main isoforms detected by isoelectric focusing. The ability of Fur to adopt oligomeric structures was assessed by chemical cross-linking, native PAGE, and MALDI-TOF experiments (not shown). Our data suggest that, in solution, Fur exists in several discrete oligomeric species, namely monomers, dimers, trimers, tetramers and even higher oligomeric forms. The quaternary structure of Fur may be of physiological importance and its study will help to understand its role in vivo, with different promoter targets or to increase affinity for the iron-boxes. Metals promoting the oligomerization could have some physiological significance relating the role of iron as corepressor.

FIGURE 3: Native PAGE of purified Fur protein at pH 4.5, showing the oligomerization of Fur after storage. Lane 1: 8 μg of Fur after storage at a final concentration of 70 μM, Lane 2: 10 μg after storage at 600 μM. 10 μg of myoglobin were used as a marker (lane 3)

Analysis Of Phosphorylation of FurA In *Anabaena* sp. PCC 7120

Phosphorylation is also a common way to regulate protein activity. In order to test whether FurA could be phosphorylated in vivo, we have isolated and purified native FurA from the cyanobacterium *Anabaena* PCC 7120. Presence or absence of phosphorylated serine and threonine residues has been analyzed using anti-phosphoserine, anti-phosphothreonine and anti-tyrosine antibodies. Results shown in Figure 4 indicate that *in vivo*, none of these residues are phosporylated in FurA.

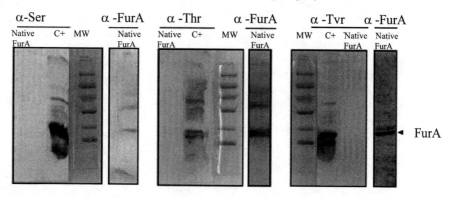

FIGURE 4. Western blot of crude extracts of *Anabaena* PCC 7120 showing absence of phosporilation. C+, positive control, MW molecular weight markers.

REFERENCES

1. S.C. Andrews, A.K. Robinson and F. Rodríguez-Quiñones, FEMS Microbiology Reviews 2003, 27, pp. 215-237.
2. K.H. Coale, K.S. Johnson, S. Fitzwater, E. Gordon, et al. Nature, 1996, 383, pp. 495-501.
3. N. Straus. The Molecular Biology of cyanobacteria, . Bryant, D. ed. Kluwer. 1994.
4. M. Kaebernick, B.A. Neilan *FEMS Microbiol. Ecol.* 2001, 35, pp.1-9.
5. J.A. Hernández, S., López-Gomollón, M.T. Bes, M. F., Fillat y M.L. Peleato *FEMS Microbiol. Letters* 2004, 236 pp. 275-282.
6. L. Escolar, J. Perez-Martin and V. de Lorenzo. *J Mol Biol.* 1998, 283, pp.537-47.
7. J.A. Hernández, S. López-Gomollón, A. Muro-Pastor, A. Valladares, M.T. Bes, M.L. Peleato and M.F. Fillat. *BioMetals*, en prensa.
8. J.A. Hernandez, M.T., Bes, M. F., Fillat, J.L. Neira and M.L. Peleato *Biochemical Journal.* 2002, 366 pp. 315-322.
9. J.A. Hernandez, A. Muro-Pastor, E. Flores, M.T. Bes, M.L. Peleato and M.F. Fillat. *J Mol Biol.* 2006, 355 pp.325-34.
10. J.A. Hernández, M.L. Peleato, M.F. Fillat and M.T. Bes *FEBS Letters*, 2004, 557 pp. 35-41.
11. C. Wyman, I. Rombel, A.K. North, C. Bustamente and S. Kustu *Science* 1997, 275 pp. 1658-1661.

Analysis of Apoflavodoxin Folding Behavior with Elastic Network Models

M. Cotallo-Abán*, D. Prada-Gracia†,*, J.J. Mazo†,*, P. Bruscolini*, F. Falo†,* and J. Sancho**,*

*Instituto de Biocomputación y Física de Sistemas Complejos. Universidad de Zaragoza. SPAIN
†Dpto. Física de la Materia Condensada. Universidad de Zaragoza. SPAIN
**Dpto. Bioquímica y Biología Molecular y Celular. Universidad de Zaragoza. SPAIN

Abstract.
We apply simple elastic network models to study some properties of the unfolding of apoflavo-doxin, a protein that shows a three-state thermodynamic behavior under thermal denaturation, as revealed by extensive analysis of wildtype and mutant variants. The intermediate of apoflavodoxin presents an overall structured core, with just a part of the protein being substantially unfolded [1]. In agreement with these results, we have been able to identify, using different models and methods, the more mobile regions in the thermal unfolding of the protein. We also discuss how the predictions obtained from these models could help in designing new experiments.

Keywords: protein, gaussian network model, statistical mechanics simple models
PACS: 87.15.Aa,64.60.Cn

THE APOFLAVODOXIN FROM ANABAENA

Flavodoxin is a 169 residue-long protein involved in electron transfer processes in Anabaena PCC 7119 and many other procaryots. Its "apo" form (1FTG), which lacks the FMN cofactor, shows a three-state thermodynamic equilibrium behavior under thermal denaturation [2, 1]. In the intermediate, a large part of the protein remains close to the native fold, but there is a non-contiguous 40-residue region which appears unfolded.

Experiments suggest that the apoflavodoxin thermal intermediate, which appears at 317.3 K, is mainly formed by the packing of helices and β strands. In contrast there are three loops quite weakened. The most significant regions are loops 57-60, 90-100, that bind the FMN cofactor and 120-139, which contains a three-stranded β-sheet [2, 1].

These results, were obtained by performing equilibrium ϕ-analysis [1], which does not allow to "see" directly the structure of the intermediate state, which, due to the intrinsic technical difficulties, has not been crystallized, nor characterized by NMR. The hypothetical structure of the intermediate has been deduced thanks to the interpretation of the changes in the relevant and residual stability of the protein, which allows to assess if a mutation changing the stability of the native state also affects the stability of the intermediate state, with respect to the unfolded state.

For this reason, in order to get a better insight on the behavior of the system, it can be useful to obtain results using different theoretical models, see if such results support the standard interpretation, and look for suggestions of new experiments that can further confirm this view.

CP851, *From Physics to Biology; BIFI 2006 II International Congress,*
edited by J. Clemente-Gallardo, Y. Moreno, J. F. Sáenz Lorenzo, and A. Velázquez-Campoy

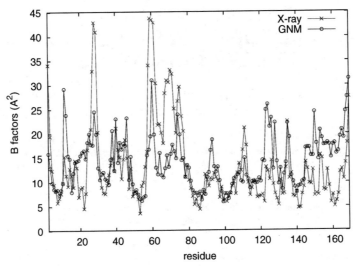

FIGURE 1. Bfactors calculated from the GNM (scaled with a factor) versus the experimental Bfactors

The simplest model we have used is the so-called "Gaussian Network Model" (GNM) which only considers the topology of the native state. We start studying its standard version [3, 4, 5], which allows us to predict the active regions of the protein (that coincide with the most mobile ones). Moreover, conformational motions of the native state, as well as coupled movements between regions, can be detected. Then the characteristics of the dynamics of the relevant protein regions is understood.

Next, we study an extended version of the GNM, introduced by Micheletti and coworkers [6, 7]. This version allows for breaking of native contacts, thus driving the protein to thermal unfolding. We deal with this model in two very different ways: through a self-consistent approximation (an analytical method assisted by numerical calculation) and by molecular dynamics simulations using a Langevin bath (numerical simulation of the dynamics of the protein).

Finally, we compare the results with all-atom simulations of the protein dynamics at two different bath temperatures. We present a discussion of our preliminary results in the light of their capability to support the three-state model for apoflavodoxin thermal unfolding.

MODELS: TWO STATISTICAL-MECHANICS APPROACHES

A.- Gaussian Network Model

As a first approach, we have studied the behaviour of protein as an elastic network [4]. Recently, this kind of models have been developed to obtain information about the mechanical properties of the native conformation [3, 4, 5].This model will give us

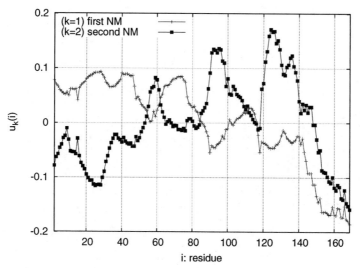

FIGURE 2. First and second normal mode

insight on the elastic properties of the crystallized structure. The approximation is as follows: the protein is reduced to a set of nodes linked with their neighbours which are at a distance less than a given cutoff (r_c). In this approximation, it is straightforward the calculation of the normal modes, cooperative motions, and the correlation of the fluctuation of the nodes. The more contribution of a set of nodes in the lowest frequency modes, the more flexible is the region. Here, we will focus on the two low frequency modes. Our interest is on the flexibility of the loops that will probably lead to the intermediate state.

The Kirchoff matrix,Γ, obtained from the crystallized structure, is build as follows: $\Gamma_{i,j} = -1$ if $i \neq j$ and $R_{i,j} \leq r_c$, 0 if $i \neq j$ and $R_{i,j} > r_c$ and $-\sum_{i,j \neq i} \Gamma_{i,j}$ if $i = j$. We define R_i as the position of the C_α atom of residue i and $R_{i,j}$ represents the distance between C_α's i and j in the crystallized structure.

From this matrix the correlation of the fluctuations around minima can be extracted:

$$\Gamma = U \Lambda U^T \qquad \Gamma^{-1} = \sum_{k=2}^{N} \lambda_k^{-1} \vec{u}_k \vec{u}_k^T \qquad \left\langle \Delta R_i \cdot \Delta R_j \right\rangle = \frac{K_B T}{\gamma} \Gamma_{i,j}^{-1}, \qquad (1)$$

being $\vec{u}_k$ the k-th column of U (that is, k-th eigenvector of Γ). This column is proportional to the k-th normal mode of the system. Λ is the diagonal matrix of eigenvalues λ_k, where:

$$\lambda_1 = 0 < \lambda_2 < ... < \lambda_n . \qquad (2)$$

Mean square deviations of each C_α in the model and the experimental Debye-Waller (B_i) factors measured by X-ray difraction are related with the equation

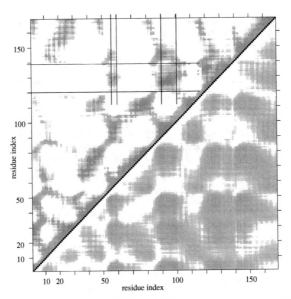

FIGURE 3. The values of $\langle \Delta R_i \cdot \Delta R_j \rangle / \langle \Delta R_i^2 \rangle^{1/2} \langle \Delta R_j^2 \rangle^{1/2}$ extracted from the GNM. The upper diagonal pixels represents the positive values, the darker the more correlated residues. Below the diagonal the negative values are represented.

$$B_i^{GNM} = \frac{8\pi^2}{3} \langle \Delta R_i \cdot \Delta R_i \rangle = \frac{8\pi^2 K_B T}{3\gamma} \Gamma_{i,i}^{-1} \, , \tag{3}$$

where γ is a scale parameter to be fitted. The other free parameter is r_c. Best fit (see Figure 1) has been achieved with $r_c = 7$ Å which is a value in agreement with those ones used in literature [4].

Figure 2 plots the first and the second modes, $\vec{u}_1$ and $\vec{u}_2$. The first one corresponds to a breathing motion, where half of the structure move against the other half. This mode does not present interesting structural features. The zones with less contribution are in the core or are playing the role of "elbow" in hinge motions. The second mode is more interesting because four relevant regions can be identified. The three loops (57-60, 90-100, 120-139) are correlated between them and uncorrelated with the region around 20-30, taking part in this way on a collective motion which represent an important part of the total fluctuation.

Normalized correlations are shown in Figure 3. A high correlation can be observed between the regions of the protein we are interested. In fact, we can see in the figure how the 120-139 loop is correlated with the loops 57-60 and 90-100. This corroborates the observations given in the second normal mode.

In conclusion, we see how this model, based in the protein topology, reveals a cooperativity between regions which can take part in the thermal intermediate. This approach allows us, as well, to do the normal modes analysis in a less time consuming way than using an all-atom simulation program. These normal modes deliver important information about long range cooperative effects. In our case, these results are a good starting point to check the hypothetical intermediate structure deduced from the experimental work with theoretical models.

B.- Extended GNM with contact breaking: self-consistent approximation

Now, we introduce the possibility to break contacts, and study the effect of temperature on the loops unfolding. To this end, we consider the Hamiltonian [6, 7]:

$$\tilde{\mathscr{H}} = \frac{T}{2} K \sum_{i=1}^{N-1} (\vec{r}_{i,i+1} - \vec{r}^{0}_{i,i+1})^2 + \frac{1}{2} \sum_{i \neq j} \varepsilon_{i,j} \Delta_{i,j} \left[(\vec{r}_{i,j} - \vec{r}^{0}_{i,j})^2 - R^2 \right] \theta_{i,j} , \qquad (4)$$

where $\theta_{i,j} = \Theta\left(R^2 - (\vec{r}_{i,j} - \vec{r}^{0}_{i,j})^2 \right)$ (the Heaviside function), $\vec{r}_{i,j} = \vec{r}_i - \vec{r}_j$ and $\vec{r}^{0}_{i,j}$ the same for the native conformation. $\Delta_{i,j}$ is the native contact map for the protein ($\Delta_{i,j} = 1$ if residues i and j are in contact in the native structure (in our case: at a distance less than 6.5 Å), $\Delta_{i,j} = 0$ otherwise).

As a first approach, we resort to a self-consistent approximation to evaluate analytically the partition function and all relevant thermodynamical quantities.

Upon replacing $\theta_{i,j}$ by a parameter $p_{i,j}$ the Hamiltonian turns into a sum of quadratic terms, which allows analytical integration of the partition function:

$$\begin{aligned}
Z(T) &= \exp\left[\frac{R^2}{2T} \sum_{i,j} \Delta_{i,j} p_{i,j} \right] \int \prod_{i=1}^{N} d^3 r_i \exp\left[-\frac{1}{2} \sum_{i,j} \vec{x}_i M_{i,j} \vec{x}_j \right] \\
&= N^{-\frac{2}{3}} (2\pi)^{\frac{3(N-1)}{2}} (\det{}' A)^{-\frac{3}{2}} ,
\end{aligned} \qquad (5)$$

where (') means that we are calculating the determinant of the matrix without considering the eigenvalue $\lambda_1 = 0$ (that is, $\det{}' A = \prod_{i=2}^{N} \lambda_i$).

From the expression above, all the relevant thermodynamical quantities can be evaluated: we have, for the average energy:

$$\langle E \rangle = \frac{3(N-1)T}{2} - \frac{R^2}{2} \sum_{i,j} \Delta_{i,j} p_{i,j}(T) , \qquad (6)$$

and the average number of native contacts or "native-state-overlap":

$$Q = \frac{\sum_{i<j} \Delta_{i,j} p_{i,j}}{\sum_{i<j} \Delta_{i,j}} . \qquad (7)$$

The parameter $p_{i,j}$ must be evaluated self-consistently, that is:

$$p_{i,j} = \left\langle \Theta\left(R^2 - (\vec{r}_{i,j} - \vec{r}^{0}_{i,j})^2 \right) \right\rangle_0 , \qquad (8)$$

where the average on the right depends on $p_{i,j}$. This average can be calculated by the integral:

$$p_{i,j} = \frac{1}{Z} \int \prod_{i=1}^{N} d^3 r_i \exp\left(-\beta H_0\right) \Theta(R^2 - (\vec{r}_{i,j} - \vec{r}_{i,j}^0)^2) , \qquad (9)$$

where $\beta = 1/T$ and H_0 is obtained from Eq. (4) upon substituting the functions $\theta_{i,j}$ with the parameters $p_{i,j}$.

By defining $\vec{x}_i = \vec{r}_i - \vec{r}_i^0$, this expression results in:

$$p_{i,j} = \frac{1}{Z} \int \prod_{i=1}^{N} d^3 r_i \Theta(R^2 - (\vec{x}_i - \vec{x}_j)^2) \exp\left[\frac{R^2}{2T} \sum_{i,j} \Delta_{i,j}\right] \exp\left[-\frac{1}{2} \sum_{i,j} \vec{x}_i M_{i,j} \vec{x}_j\right] , \qquad (10)$$

with the M the matrix defined by

$$(M)_{i,j} = \delta_{i,j}\left(K(2 - \delta_{i,1} - \delta_{i,N}) + \frac{2}{T}\sum_l \Delta_{i,l} p_{i,l}\right) +$$
$$+ (1 - \delta_{i,j})\left(-K(\delta_{j,i+1} + \delta_{j,i-1}) - \frac{2}{T}\Delta_{i,j} p_{i,j}\right) . \qquad (11)$$

The integral can be calculated by a Laplace Transform. By using R^2 as the variable t, we need to calculate:

$$\mathscr{L}(f(t)) = \mathscr{L}\left[\int \prod_{i=1}^{N} d^3 x_i \,\Theta(t - (\vec{x}_i - \vec{x}_j)^2)\exp\left[-\frac{1}{2}\sum_{i,j}\vec{x}_i M_{i,j}\vec{x}_j\right]\right] , \qquad (12)$$

and developing the calculation:

$$\mathscr{L}(f(t)) = \frac{1}{s}\int \prod_{i=1}^{N} d^3 x_i \exp\left[-\frac{1}{2}\sum_{i,j}\vec{x}_i Q_{i,j}\vec{x}_j\right] , \qquad (13)$$

with the new matrix $Q_{k,l} = M_{k,l} + 2(\delta_{k,l}(\delta_{k,i} + \delta_{k,j}) - \delta_{k,i}\delta_{l,j} - \delta_{l,i}\delta_{k,j})s$.

Now the integral is reduced to the same form as in the partition function, above. The problem with matrix Q and M is that, due to the translational invariance $\vec{x}_i \rightarrow \vec{x}_i + \vec{a}$ of the problem, the sum of the elements on each of their rows and columns is zero: this fact yields a null determinant, producing a singularity in the evaluation of the averages.

We can follow McCammon and coworkers [8] to solve the problem by considering and extra spring on the terminal residues N: we add an extra term $\frac{1}{2}\gamma \vec{x}_N^2$ to the hamiltonian, that preserves the structure of the quadratic form of the hamiltonian, and hence the form of the matrix M, but removes the translational invariance. The extra contribution to the free energy can be explicitly calculated.

With this approach, after performing the gaussian integral and the inverse Laplace transform, and restoring the variable R, the explicit expression of the $p_{i,j}$ reads:

$$p_{i,j} = -\frac{2}{\sqrt{\pi}}\rho_{i,j}e^{-\rho_{i,j}^2} + \text{Erf}(\rho_{i,j}) \,, \tag{14}$$

with:

$$\rho_{i,j} = \frac{R}{\sqrt{2\gamma_{i,j}}} \qquad \gamma_{i,j} = ((M_N^N)^{-1})_{i,i} + ((M_N^N)^{-1})_{j,j} - 2((M_N^N)^{-1})_{i,j} \,, \tag{15}$$

where $(M_N^N)^{-1})_{i,j}$ is the (i,j)-element of the inverse of the matrix obtained from M upon elimination of the Nth row and column. At any temperature $p_{i,j}$ will be found iteratively from the above expression, as described in [6].

We follow [6] in the choice of the parameters: $R = 3$, $K = 1/15$. The native contact probability per residue Q_i are defined:

$$Q_i = \frac{\sum_j \Delta_{ij} p_{ij}}{\sum_j \Delta_{ij}} \,. \tag{16}$$

In this study, we first set $\varepsilon_{ij} = \varepsilon$ for each i and j, and focus just on the geometry of the apoflavodoxin native state: we aim at understanding how much the folding geometry is responsible for the thermodynamics of the folding process, and how "trivial" the resulting thermodynamic behavior appears, as compared to that of a random contact-map.

To this end, we study the behavior of the model with several different contact maps: the original, wild-type one, and many others, obtained by random reshuffling of each residue's contacts, in such a way that the connectivity of each residue (i.e., the number of contacts it makes) is preserved, but the resultant geometry is completely random. This most likely produces non-physical contact-maps, that violate geometric constraints, but due to the nature of the model, where just deviations $\vec{x}_j$ from the equilibrium position are relevant, the value of the thermodynamical information coming from such maps is not affected, and we are allowed to compare the thermodynamic behavior of the different cases.

Then, we perturb the models introducing a little differentiation among contacts, choosing a subset and making them weaker, with their energy being 0.1ε: we want to test in this way how the introduction of a "sequence" changes the results obtained for the homogeneous case.

We choose the contacts to be weakened with three different patterns:

1. all contacts of residues in the region 94 - 124 (corresponding to a helix placed in the surface of the native wild-type protein)
2. randomly chosen contacts (irrespective of the residues involved in the contact)
3. all contacts of residues in the region 57-60, 90-100, 120-139, that correspond to the unfolded part of the thermal intermediate, according to experiments.

In the first case, there were 263 contacts, in the second one, 271 and in the last one, 243 contacts in real maps, and 285 contacts in the first case, 271 contacts in the second one and 291 contacts in the last one for the reshuffled maps, out of a total of 649 contacts

141

in all cases. We performed the study with several samples of random contacts maps, obtaining similar results: those reported in the following are typical results.

In Figure 4, 5 and 6 we observe energy, specific heath and average number of native contacts (Eq. (7))for different groups of results.

Several comments are in order:

1. We cannot find important differences in the thermodynamic behavior obtained with the original or the reshuffled map, in the homogeneous case of same ε. Even if a quantitative van t'Hoff analysis cannot be carried on in this case, due to the nature of the self-consistent approximation, if we still consider the ratio of the height by the width of the peak of specific heat as a reasonable measure of cooperativity, we see that the reshuffled and original map show practically the same degree of cooperativity.

2. Things do not change if we weaken the same number of randomly-chosen contact in the wild type and reshuffled geometry. As expected, the peak moves towards lower temperatures (due to overall decrease of stability of the folded conformation) and is somewhat shorter, but native and random geometry still produce almost identical traces in all figures.

3. In the above two cases, a single peak is present in the specific heat, suggesting a two-state behavior. This agrees with the simulations performed for the wild type geometry in the homogeneous case (see following section). Things start changing when we weaken all the contacts pertaining to a group of residues: we can see that the specific heat traces of wild type and random geometry becomes increasingly different as we move to weakening the region of the superficial helix 94-124 and then to weakening the three experimentally relevant regions 57-60, 90-100, 120-139. Moreover, the specific heats obtained with wild-type geometry starts showing a small shoulder at low temperatures in the "helix" case, that becomes a clear peak when the three different regions are weakened.

Thus, it seems that according to the geometry of apoflavodoxin, the model suggests a two-state behavior. A three-state behavior only appears when we introduce different energies for the contacts, roughly mimicking the energetic heterogeneity involved by the sequence, and arrange the weak contacts in such a way that all the contacts of a group of residues are weak. This supports the view that the specific sequence of wild type apoflavodoxin has a central role for the existence of an equilibrium intermediate, a view which agrees with the observation that the interface, between the three experimentally-determined unstable regions and the bulk of the protein, is unusually polar.

In Figure 7 we can see, for different temperatures, the average number of contacts of residue i that are still formed, for each i (Eq. (16)), for the homogeneous wild-type case. We can observe that, also at the transition temperature, when on average half of the contact are formed, the fluctuation around the average are not very pronounced, due to the "mean-field" nature of the self-consistent approximation. However, it is possible to notice that the experimentally relevant regions correspond indeed to regions that are predicted to be highly unfolded (low Q_i). This is especially true for loops 57 - 60 and 120 - 139, while loop 90 - 100 is not as well represented. It should be noticed, though, that these results are not sufficient to perform a safe prediction, on their basis, of the structure

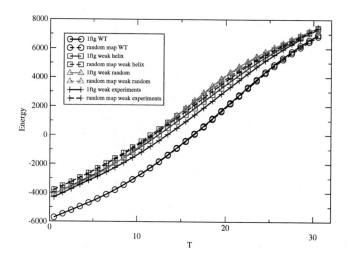

FIGURE 4. E versus T for different weakening patterns. In the legend, "weak helix", "weak random" and "weak experiments" indicate that the choice of the weak contacts is made according to pattern 1, 2 or 3 in the text, respectively.

of the intermediate. This could be expected, in the light of the above results, since we have seen that the introduction of energetic heterogeneity is necessary to reproduce three-state behavior.

C.- Extended GNM with contact breaking: Langevin dynamics

In this section, we study the Langevin dynamics of the extended GNM model described by equation (4). This study will allow us to validate the results previously obtained and to explore the space of configurations in a more realistic way. Within other important things, the dynamics will show us the manner the contacts break when the system is embedded in a stochastic thermal reservoir at temperature T.

The equations of the motion read as:

$$\dot{q}_i(t) = p_{q,i}/m \quad ; \quad \dot{p}_{q,i}(t) = -\frac{\partial H}{\partial q_i} - \gamma p_{q,i}(t) + \eta_{q,i}(t) , \tag{17}$$

where i is the C_α index and $q = x, y, z$.

In these equations, we control the temperature trough the stochastic term $\eta_{q,i}(t)$ which is a Gaussian distributed δ-correlated random noise:

$$\langle \eta_{q,i}(t)\eta_{q,j}(0) \rangle = 2\gamma m k_B T \delta(t)\delta_{i,j} , \tag{18}$$

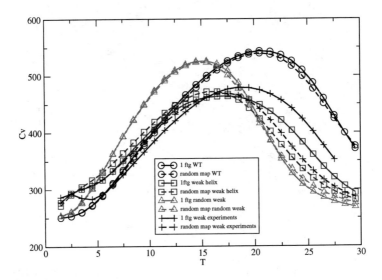

FIGURE 5. C_v versus T for different weakening patterns. In the legend, "weak helix", "random weak" and "weak experiments" indicate that the choice of the weak contacts is made according to pattern 1, 2 or 3 in the text, respectively.

where, k_B denotes the Boltzmann constant and T is the temperature (for the sake of simplicity, both k_B and m have been set to unity).

Along a trajectory, we have computed the values of some of the variables previously defined: the native-state-overlap $Q(T)$ (Eq. (7)), or the native contact probability per residue $Q_i(T)$ (Eq. (16)) for instance.

In Figure 8, we can see the value of Q_i for each residue in the protein. Once more, it is shown that the regions more amenable to broke their contacts are those which are found in experiments; that is, loops 57 - 60, 90 - 100 and 120-139.

One should note that temperature for unfolding obtained in Langevin simulation is much lower than that calculated in the self-consistent method. This is due to the fact that, since self-consistency introduce a kind of mean field approximation, it can not take into account the large fluctuations observed in the simulations. Indeed, within self-consistent approximation the contacts are never actually broken, except in the case when the corresponding p_{ij} is strictly zero: therefore, also in the denatured conformations, there is a small quadratic bias toward the native conformation. This results in the need of a higher temperature to produce the same degree of unfolding that we get in simulations of the original model.

Hovever, if we consider the results at the respective transition temperatures, we can establish a correlation between the results found with these simulations and with self-

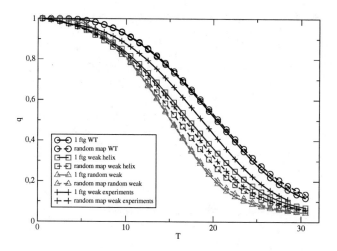

FIGURE 6. q versus T for different weakening patterns. In the legend, "weak helix", "random weak" and "weak experiments" indicate that the choice of the weak contacts is made according to pattern 1, 2 or 3 in the text, respectively.

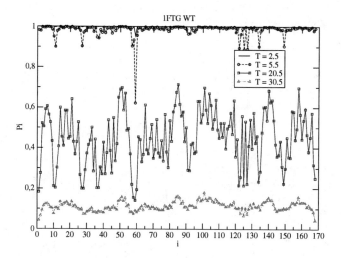

FIGURE 7. Q_i from the WT protein for different temperatures.

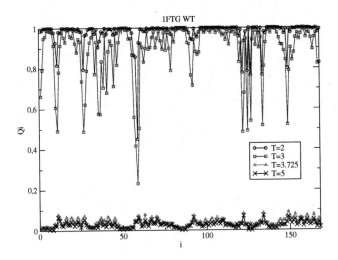

FIGURE 8. T1 = 1, with Q = 0.977133; T2 = 2, with Q = 0.890280; T3= 3.725, with Q = 0.037224; T4 = 5, with Q = 0.028062.Once more, we can see that loops, especially 57-60, are more prone to break their contacts.

consistent approximation made in previous subsection.

In order to find it, we use Pearson correlation coefficient. If we have two sets of N data, x_i and y_i, then the Pearson correlation coefficient is calculated as:

$$r_{x,y} = \frac{\sum_i (x_i - \bar{x})(y_i - \bar{y})}{(N-1)s_x s_y},$$

(19)

where $\bar{x}$ and $\bar{y}$ are the sample means of each of the data sets, and s_x and s_y are the standard deviations.

The coefficient can take values from -1 and 1. Conventionally, absolute value of correlation less than 0.1 indicates no correlation. The correlation is small between 0.1 and 0.29, medium from 0.3 to 0.49, and large above 0.5.

We want also to mention that for the sets of data corresponding to self-consistent calculations and Langevin dynamics at the respective transition temperatures, the value of the correlation coefficient is 0.97, indicating that the self-consistent approximation correctly grasps the relevant physics of the model, despite the difference in the numeric values.

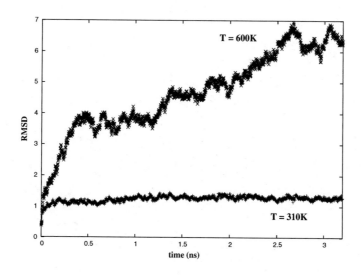

FIGURE 9. RMSD for $T = 310$ and $T = 600$ along the trajectory of the all-atom Langevin dynamics simulations

D.- All-atoms simulation

Finally, we analyze the unfolding of the protein using a complete description of the structure by mean of all-atom molecular dynamics (MD) simulations. Since this technique is computationally expensive, we restrict ourselves to only two temperatures.

Simulations were performed using NAMD[9] on a 16 parallel processor cluster. A CHARMM27[10] force field was used, with a cutoff of 12Å for non-bonded interactions. Protein was embedded in a water sphere of 31Å, large enough to avoid edge effects along the simulation run. Langevin dynamics with a friction coefficient of 5 ps^{-1} was run up to 3 ns. Two different temperatures, 310K and 600K, were simulated. The first one, in order to check the stability of the protein with the simulation parameter used. The second, to induce a fast thermal unfolding and to observe the pathway followed in the process.

After simulation we extract the global RMSD (figure 9) and the RMSD per residue (figure 10). Figure 9 shows the fast unfolding at high temperature whereas the native structure remains stable giving confidence to the simulations. In figure 10 we observe the residues that contribute mostly to RMSD. From the maxima and minima of RMSD of each residue, we corroborate that previous methods identifies more active and less active zones. In agreement with the previous findings, the loops are the most mobile zones, and the first to loose their native structure when we perform a thermal denaturation.

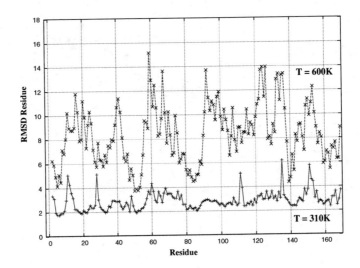

FIGURE 10. RMSD per residue for $T = 310$ and $T = 600$ at the end of all-atom Langevin dynamics simulations

CONCLUSION

We have used two mesoscopic network models to make a statistical mechanics study of the apoflavodoxin unfolding. Within the GNM framework, we have obtained regions which are correlated in their motions. This analysis reveals that loops 57-60, 90-100 and 120-139 are the more flexible protein regions. Using the extended GNM with contact breaking we have obtained the thermal behavior of the unfolding process. This model has been studied with two approaches: a self-consistent method and the Langevin dynamics of the same system. With both methods, we find that the residues that break first the native contacts correspond to those which present high flexibility in the GNM analysis. Finally, preliminary all-atoms MD simulations supports the above findings .

At present, we can neither precise the structure of the intermediate nor support with evidence the hypothetical intermediate proposed in [1]. To this end, further refinement of the model, including the introduction of sequence heterogeneity, is necessary. Such work is under present development. However, our results point in the right direction, allowing to mark the regions in the protein candidates to be broken in the equilibrium intermediate, and possibly suggesting new experiments focused on other unexplored regions.

REFERENCES

1. L. A. Campos, M. Bueno,J. Lopez-Llano,M. A. Jimenez, and J. Sancho,"Structure of Stable Protein Folding Intermediates by Equilibrium phi-Analysis: The apoflavodoxin Thermal Intermediate." *J. Mol. Biol* **344**, (2004)
2. M. P. Irun, M. M. Garcia-Mira, J. M. Sanchez-Ruiz, and J. Sancho, "Native hydrogen bonds in a molten globule: the apoflavodoxin termal intermediate", *J. Mol. Biol* **306** (2001)
3. I. Bahar,A. R. Atilgan, M. C. Demirel, and B. Erman, *Phys. Rev. Lett.* **80**, (1998).
4. T. Haliloglu,I. Bahar, and B. Erman, *Phys. Rev. Lett.* **79**, 3090-3093 (1997).
5. I. Bahar, B. Erman,R. L. Jernigan, A. R. Atilgan, and D. G. Covell *J. Mol. Biol.* **285**, 1023-1037 (1999).
6. C. Micheletti, J. R. Banavar, and A. Maritan, *Phys. Rev. Lett.* **87**, 88102 (2001).
7. C. Micheletti, F. Cecconi, A. Flammini, and A. Maritan, *Protein Science* **11**, 1878-1887 (2002).
8. T. Shen, L. Canino, and J. McCammon, "Unfolding proteins under external forces: a solvable model under the self-consistent pair contact probability aproximation", *Phys. Rev. Lett.* **89** 2002
9. J. C. Phillips, R. Braun, W. Wang, J. Gumbart, E. Tajkhorshid, E. Villa, C. Chipot, R. D. Skeel, L. Kale, and K. Schulten. *J. Comp. Chem.*, **26**, 1781-1802, (2005).
10. B. R. Brooks, R. E. Bruccoleri, B. D. Olafson, D. J. States, S. Swaminathan, and M. Karplus. *J. Comp. Chem.*, **4**, 187-217 (1983)

Current trends in the modeling of biological networks

Yamir Moreno*, Luis Mario Floría[†,*] and Jesús Gómez-Gardeñes[†,*]

*Instituto de Biocomputación y Física de los Sistemas Complejos.
Universidad de Zaragoza, E-50009 Spain.
†Departamento de Física de la Materia Condensada.
Universidad de Zaragoza, E-50009 Spain.

Abstract.
How does the interplay between complex structures and nonlinear dynamics may shed new light on what is going on at the cellular and molecular levels of organization of biological systems? As in other natural systems, on one hand, scientists have begun to look for patterns of interactions in biological systems. The idea behind this approach is that we can not completely understand the functioning of the cell by studying its components separately. The next step consists of taking into account the dynamics governing the unraveled interactions. This is certainly not an easy task as one has to deal with two sources of complexity: one coming from the unraveled structural patterns and the other from a dynamics in which analytical insights are difficult to take. Here we summarize some of the most recent and important works addressing the network approach to biological systems.

Keywords: Biological Networks, Boolean Dynamics, Power-law Distributions, Michaelis Menten Dynamics
PACS: 89.75.-k; 95.10.Fh; 89.75.Fb; 05.45.-a

INTRODUCTION

In 1999, Hartwell and collaborators published an influential paper discussing the new challenges of modern biology [1]. The authors pointed out that an issue of utmost importance is to develop a general framework in which biological functions could be understood as part of a complex modular organization of molecules or cell's constituents. In other words, modern biology should explain not only the functioning of individual cellular components, but also how these components are interconnected through a complex web of interactions leading to the function of a living cell. It is then natural to ask what these biological networks at the cell organization level look like and how their structure couples to the dynamics.

Cells are life's fundamental units of structure and function. It was expected that, once the complete instructions encoded in DNA would have been interpreted, one could map a gene (the basic information unit in the DNA) into a specific activity or function, with all the consequent potential applications such as targeted drug development [2]. On the contrary, although today the complete knowledge on the genes of several organisms is available, yet the relationship between blueprints in DNA and functional activities of the cell is not fully understood. For instance, the p53 gene and protein (having the function of controlling cell's life and death) are known as tumor-suppressor, since it was found that the p53 protein does not function correctly in most human cancers. However,

CP851, *From Physics to Biology; BIFI 2006 II International Congress*,
edited by J. Clemente-Gallardo, Y. Moreno, J. F. Sáenz Lorenzo, and A. Velázquez-Campoy
© 2006 American Institute of Physics 0-7354-0350-3/06/$23.00

despite the many studies performed on p53 gene and protein, the way on how effectively suppressing the growth of cancer cells is missing at a genetic level. Recently, it has been proposed that the understanding of such cancer cell growth mechanism would be gathered not only from the study of the p53 gene and protein, but taking into account the whole network interacting with them [3]. That is, the function of the gene should be analyzed through a network in which the gene participates. Similarly to p53 network case, several other observations prove that some functional activities of the cell emerge from interactions between different cell's components through complex webs. Moreover, it is expected that the large-scale network approach may lead to new insights on various longstanding questions on life, such as robustness to external perturbations, adaptation to external circumstances, and even hidden underlying design principles of evolution.

In what follows, we discuss the last advances in the characterization of some biological networks from two points of view: their structural organization and their functioning. The main point here is how to uncover the relationship between the two sources of complexity intimately linked (dynamics and structure) as both play a key role in the functioning of the system. We stress here that our main intention is to provide a brief overview of the current state in the field, and that many works may be overlooked due to space constraints. We invite the interested reader to follow the specialized literature.

STRUCTURE

A plenty of cellular and molecular networks have been unraveled in the last several years. We here refer to those that have been more used in subsequent studies or because they are considered to be essential for the cell's life.

The first of these complex biological networks is that formed by metabolic reactions: the metabolic network. Jeong et al. have considered the metabolic reactions of 43 different organisms, representing the three domains of life, and have constructed directed graphs whose nodes are the metabolites and edges represent biochemical reactions [4]. A node receives an incoming edge when the corresponding metabolite is produced, and receives an outgoing edge when the metabolite is educed. Enzymes are not included in the graph. The total number of connections (edges) of a node is called the degree of the node. If the edges have a direction (incident to or going out from the node), the degree of a node is divided in in-degree and out-degree, respectively. For all investigated organisms, the resulting graphs for metabolic reactions exhibit scale-free properties for both incoming and outgoing degree distributions.

Scale-free networks refer to a class of graph in which the probability that a node is connected with k other nodes follows a power law distribution $P(k) \sim k^{-\gamma}$, with $2 \leq \gamma \leq 3$ in the vast majority of networks. This is a property not only found in biology, but in fields as diverse as social, technological and natural systems [5]. Besides, scale-free networks are characterized by what is known as *small-world property*, which means that the average distance between any two nodes of the graph scales at much as $logN$, where N is the size of the system. The power-law distribution for the degree of the vertices of a network and the small world property are common topological features of many real world networks [5].

The above-mentioned properties were found universally, irrespective of metabolic pathway databases and of the methods used to construct graphs from biochemical reactions. For example, instead of assuming virtual intermediate complexes, Wagner and Fell built up two networks (the metabolite and the reaction networks) from the metabolic pathways of *Escherichia coli* [6]. The metabolite network consists of nodes representing metabolites and bidirectional links between educt and product of a metabolic reaction. On the other hand, the reaction network is the network where the nodes correspond to metabolic reactions and two nodes are linked when the two reactions share a metabolite. In metabolite networks, scale-free properties are detected, while the reaction network does not show power-law degree distributions. Small-world properties and relatively high clustering (i.e, how probable it is that two nodes with a common neighbor are also connected together) are found in both networks. Other studies with different ways of obtaining graphs show almost identical results [7, 8, 9, 10].

Another class of well-studied cellular networks is that of protein-protein and protein-gene interaction networks. This is due to the increasing availability of data sets and new experimental techniques that allows a more detailed study of the interactions at the cellular level. On the other hand, interactions among proteins have a crucial role in several functional activities, such as signal transduction. According to the demand of understanding the protein interaction map, several high-throughput experiments have been performed. They provide evidence of a partial interaction map between proteins. In the graph representation, a node corresponds to a protein and two proteins are linked when they physically interact. The east two-hybrid screen method has been applied for revealing protein-protein interactions by Uetz et al. [11] and by Ito et al.[12]. Similarly to metabolic networks, scale-free properties, high-clustering and small-world properties have been found. Besides, the studies performed have allowed to address other questions such as the robustness of these networks against random and directed failures [13]. It should be noticed that the databases used in the analysis show very small overlap, while the individual networks obtained from each database show a very similar structure. In particular, it has been argued that the biological functional organization and the spatial cellular organization are correlated significantly with the topology of the network, by comparing the connectivity structure with that of randomized networks.

Finally, we note that networks constructed from gene expression data are currently under exploration [14, 15]. For instance, Agrawal [15] have studied networks from gene expression of cancer data. By analyzing individual gene expression level at different samples, networks in which the degree distribution of the nodes shows a power-law behavior in the tails with an exponent 1 can be constructed. Stuart et al. have further shown that co-expressed gene networks of humans, flies, worms, and yeast have scale-free properties [14].

In summary, biological networks seems to share many topological properties. What do these properties mean in a biological system? And what basic principles in biology give rise to such universal features? Many steps toward the answers to these questions have been certainly given in the last several years. However, the majority of the issues addressed are based mainly on analyzing the structure of these networks without taking into account their dynamics, i.e., the fact that the structure correlates with the functioning of the underlying system. For instance, from a topological point of view, it has been argued that the nodes with a high degree (the hubs, those contributing to the tail of the degree distribution) are critical for the robustness of the system to intentional removal of them. On the other hand, the hubs have been shown to radically change the behavior of the system in front of several dynamical processes such as epidemic spreading [5, 16]. It is yet to see whether or not the same results hold when nonlinear dynamics coexists with complex topological structure. We next describe two promising approaches in this direction.

DYNAMICS

During the last several years a wealth of experimental data, obtained with technological advances such as cDNA microarrays, have allowed the dynamical characterization of several biological processes both on a genome-wide and on a multi-gene scales and with fine time resolution. From a theoretical side, compelling models on the dynamics governing metabolic and genetic processes are hard to build as these biological phenomena are highly nonlinear and with many degrees of freedom. However, scientists have certainly advanced towards a comprehensive global understanding of, for instance, gene regulation through genetic engineering that require a thorough understanding of the general principles that can guide the design process. It is impossible here to provide an exhaustive review of the subject. However, we think that it is important to provide at least some ideas about the research lines that relate the structure and the function of biological systems.

Concepts such as operon, regulator gene and transcriptional repression were first introduced in the literature by Jacob and Monod [17]. Their model has served as the basis for more elaborated models as different regulatory mechanisms have been discovered [18]. Recent theoretical studies capitalize on these kind of models in order to elucidate what are the system constituents, their properties and how they interact in order to give rise to the collective behavior of the system. The final goal is to understand the relationship between structure and function as determined by the biological environment. In this sense, different gene circuit designs should be compared to determine which of them confers selective advantage in an ecological context and thus one should be able to advance what the functional consequences of different designs are. This is usually done by exploring the parameter space and looking for performance criteria such as the ability of a system to return to a steady state after a perturbation (called stability) or its responsiveness, that can be measured as the recovery time of the system after an environmental change (a change in an independent variable).

The results obtained for elementary gene circuits certainly provide answers to intriguing questions about how gene circuits could be organized, but at the same time pose new ones. With the recent advances in the characterization of the structure of gene networks, it is clear that genome-wide approaches will allow to discover new higher-order patterns. Therefore, more efforts in modeling the dynamics of increasingly complex gene circuits are expected in the near future. Some steps in this direction have been given.

Boolean modeling of regulatory networks

The first attempt to describe the functioning of genetic regulatory networks was performed by S.A. Kauffman [19]. This pioneering work settled the basis for modeling the complex nature of dynamics and interactions between genes and their products. In his work, each gene, i, and its product, I, were abstracted as a node of a random network having a fixed number, k, of neighbors that regulate its level of activation, g_i. This level of activation can be viewed as the concentration of the transcribed mRNA and/or the protein I encoded. The boolean character of the formulation done by Kauffman implies a qualitative description of whether a gene is activated ($g_i = 1$) or not ($g_i = 0$). Besides, time is considered as a discrete variable so that the dynamical behavior of the gene ensemble is described by the temporal series of their activity levels. At each time step the activity level of a single gene is updated considering the state of its k neighbors

$$g_i(t + \tau) = f_i(g_{j_1}(t), ..., g_{j_k}(t)) . \tag{1}$$

This is performed by means of booleans functions, f_i, that make use of the basic "AND", "OR" and "NOT" logical functions so that the results can be either 1 if the statement is true or 0 if it is false. The construction of each boolean function depends on the particular interactions that a gene shares with its regulators and has to be carefully analysed with the help of biochemical data. On the other hand, the work by Kauffman was performed from a general point of view and considered a random assignment of the boolean functions that governs the dynamical evolution of the gene's activity. The main result of the work is the existence of a phase transition on the number and length of the dynamical attractors. In particular, for $k > 2$ the number of cycles scales with the number of genes, N, and its length scales exponentially with N. On the other hand, for the case $k = 2$ these two quantities scale as $\sqrt{N}$. The above findings are biologically relevant if one considers that different genetic dynamics can be regarded as biologically differenciate cells. Taking into account that the cell diversity of a living organism scales approximately with the square root of the genetic population Kauffman suggested that gene regulatory networks should operate just on the border of the dynamically ordered region.

The above findings represented the starting point of a lot of research on the so-called subject of "Kauffman networks" during the last 25 years. These works mainly focus on the search of a full description of the dynamically different regions as well as the characterization of the phase transition (recent work on the matter can be found in [20, 21, 22, 23, 24, 25]). On the other hand, "Kauffman networks" have served as a framework for performing a coarse-grained description of real gene regulatory networks. The availability of real regulatory networks inferred from DNA microarray data joined with the easy implementation and management of the boolean dynamics provides a useful tool for understanding the interplay between the topology and the function of biological networks.

The use of boolean dynamics to characterize real genetic regulatory networks has been recently applied to the case of the *segment polarity genes in the Drosophila Melanogaster* [26]. In this case the whole map of interactions between genes is known and Boolean dynamics is seen to reproduce the patterns of gene expression that appear in the wild type. Besides, it has been tested when mutations are present confirming the validity of the model. The application of this method can help to determine the effects of new mutations and constitute a test for the question of whether the topological features of the interaction network or the kinetic details play the key role in the functioning of biological networks. The success of the use of Boolean modeling points out that it is the former which is the relevant ingredient. Another recent application of Boolean dynamics to a real gene circuit is found in [27] where the *yeast transcriptional network* is considered. In this case the point of view is drastically different because neither the nature of the interactions between genes nor any dynamical state of the system is available. The starting point is simply a set of connected genes and the authors apply a Boolean modeling of the interactions for determining what set of (Boolean) interaction rules lead to a stable dynamics of the whole system. The authors also study the effect of rewiring links of the network and conclude that dynamical states on top of the original network is more stable than on the perturbed ones. The above two examples show how the coarse-grained Boolean modeling can help to analyze the large amount of available experimental data and answer the question on where the biological stability observed has its roots.

Finally, let us remark that the boolean modeling can be reformulated in order to incorporate realistic features of real regulatory networks. Perhaps, the most important ingredient is to reproduce the effects of noise (which is a substantial characteristic of a biological system). This is usually incorporated on the form of a non synchronous update rule, assigning a time delay to each variable of the Boolean functions, f_i. Another interesting extension of the formulation is considering multi-levels for the gene activity so that the model incorporates some quantitative description on how much the gene is activated.

Continuous time modeling of dynamics

Saturated response

There is a wide variety of situations in which the system response to an external action saturates. Perhaps the most familiar example of saturable behavior known to physicists is the adsorption of gas molecules on a solid surface: At thermodynamical equilibrium, the fraction (coverage ratio) θ of surface interstitial occupied by adsorbed molecules depends on the gas pressure P as [29]

$$\theta = \frac{P}{P_0(T) + P} \qquad (2)$$

where the temperature-dependent constant $P_0(T)$ is the pressure value at which the coverage ratio reaches half of its possible maximal value $\theta = 1$. While for small values of P, compared to $P_0(T)$, θ increases linearly with P, for values of the pressure larger than $P_0(T)$ the coverage ratio becomes insensitive to pressure variations. Saturable behaviours of this type (and of a more general form; see below) have been introduced [30] in the modeling of interactions among species in ecological systems, where (most notably) they effectively provide robustness to the limit-cycle behaviour often observed in these systems [31, 32]. In the realm of Social Sciences, saturated response functions have been also used to model some type of social interactions like *e.g.* the effects of community investments in police pressure and/or educational programs on the street-gang growth phenomena [32].

Biological reaction rates are often saturable; while at small concentrations of a new chemical introduced in a cell, this responds sensitively, the response should not keep growing indefinitely as the new chemical concentration grows. The archetypal example of saturation in biological systems is the Michaelis-Menten equation [33] governing the concentration evolution of a product catalyzed from a substrate by an enzyme which binds to it. If x and y denote the concentrations of product and substrate respectively, then the reaction rate is given by

$$\frac{dx}{dt} = \frac{V_{max}y}{K_M + y} \tag{3}$$

where K_M is called the Michaelis constant and V_{max} is the value at which the rate saturates for high substrate concentrations. This saturation behaviour can be understood from the usual chemical kinetics (law of mass-action) in an intuitive way: when the enzyme molecules are mostly bound to substrate molecules, adding more substrate cannot speed up the reaction [34]. If n, instead of only one, substrate molecules bind to the enzyme, the reaction rate takes a more general functional form of saturation, often called Hill equation

$$\frac{dx}{dt} = \frac{V_{max}y^n}{K_M + y^n} \tag{4}$$

showing a sudden increase of the reaction rate towards saturation around $y = K_M$. The Hill parameter n often takes on non-integer values. Both Michaelis-Menten and Hill equations are often used in models of biological reactions, even when the explicit mechanisms generating them are unknown in many cases.

Synthetic genetic networks

In cells, the proteins, RNA and DNA form a complex network of interacting chemical reactions governing all cellular functional activities like metabolism, response to stimuli, reproduction, ... While the understanding of the structure of these networks is growing rapidly, the current understanding of their dynamics is still rather limited. In this regard, an interesting body of research is currently addressed to synthetic genetic networks, which offer an alternative approach aimed at providing a controlled test bed for the detailed characterization of some isolated functions of natural gene networks, and also pave the way to engineering of new cellular behaviour.

An example of synthetic gene regulatory network, termed the "repressilator", is becoming one of the best studied model systems of this kind. The repressilator is a network of three genes, whose products (proteins) inhibit the transcription of each other in a cyclic way; they were added to the bacterium *E. coli*, so periodically inducing the

156

synthesis of green fluorescent protein as a readout of the network state [35]. The authors of the work first argue that the repressilator can show temporal fluctuations in the concentration of each of its components, by analyzing a system of six ODE's (which, in turn, were obtained by a process of integration-out or coarse-grain away of the promoter states involved in the regulation, and rescaling of the variables) modeling the network. If p_i ($i = 1,2,3$) denote the three repressor-protein concentrations (in units of the Michaelis constant K_M), and m_i their corresponding mRNA concentrations (appropriately rescaled), the repressilator equations are (assuming the symmetrical case in which all three repressors are identical except for their DNA-binding specificities):

$$\frac{dm_i}{dt} = -m_i + \frac{\alpha}{1+p_i^n} + \alpha_0 \tag{5}$$

$$\frac{dp_i}{dt} = -\beta(p_i - m_i) \tag{6}$$

where $i = 1,2,3$ and $j = 3,1,2$; α_0 ($\alpha + \alpha_0$) is the number of protein copies produced from a given promoter type in the pressence (absence) of saturating amounts of repressor, β is the ratio of the protein decay rate to the mRNA decay rate, and time is rescaled in units of the mRNA lifetime. This system of equations has a unique steady state which can be stable or unstable depending on the parameter values. In the unstable region of parameter space, the three protein concentrations fluctuate periodically. Experiments show temporal oscillations of fluorescence, which were checked to be due to the repressilator. Though admittedly oversimplified, the model of ODE's guided the experimental design, for it served to identify possible classes of dynamic behaviour and to determine which experimental parameters should be adjusted in order to obtain sustained oscillations.

Not surprisingly, the repressilator called attention from experts on (biological) synchronization, for it offers good prospectives for further insights into the nature of biological rythms, whose mechanisms remain to be understood. In this respect, in reference [36] the authors propose a simple modular addition (of two proteins) to the repressilator original design, which allows for a mechanism of coupling between cells containing the repressilator networks.

Modules

As seen in the previous subsection, even a very small gene network, like the repressilator, requires some simplifying assumptions for an analysis of its dynamic behaviour in terms of ordinary differential equations. With large networks involving thousands of regulatory genes, this approach would require a huge number of differential equations and, what is even more problematic, an exploding number of dimensions of the parameter space (decay rates, production rates, interaction strengths, etc.). Thus an important issue concerns the right level of description when constructing quantitative models of large genetic networks [28].

In this regard, several works (*e.g.* [37, 38, 39]) have focussed on the identification of general building blocks (motifs) in genetic networks, showing robust or "reliable" behaviour. These include small modules of a few genes, such as autoregulatory excitatory feedback loops, inhibitory feedback loops, feed-forward loops and dual positive-feedback loops, which represent different kinds of robust switching elements, whose occurrence as subgraphs in real networks is significantly higher than in their randomized versions. These works provide support to discrete models in which genes are modeled as switchlike dynamic elements that are either "on" or "off", of the Boolean type described in the previous section, and point toward strong correlations between structural and functional properties of genetic regulation networks.

The robustness of slightly larger modules, like the segment polarity genes of the fruit fly *Drosophila* (a subgraph of the segment determination network, responsible for the embryonic development of the insect body segments), has been convincingly tested with a realistic dynamical model [40] supporting the view that segmentation is modular, with each module autonomously expressing a characteristic intrinsic behaviour in response to transient stimuli. A connectionist model for the segment determination system of *Drosophila*, including cell-cell interaction via one-dimensional diffusion [41] has been thoroughly characterized (along with its continuum limit (PDE) equations [42]). These generalised reaction-diffusion models inspired further work in [43] which identified minimal gene networks associated to different segmentation patterns; also, extensive computer simulation of randomly generated networks showed that combinations of spatial patterns can be mapped into combinations of the basic modules.

The resistance of modules to variations (proxy for mutations of small effect) in the kinetic constants and various parameters that govern its dynamical behaviour, may suggest that evolution could rearrange inputs to modules without changing their intrinsic behaviour, or as conjectured in [43], that the target of selection would operate not only on single-gene level structures, but also on the available structures in the high-dimensional parameter space of the model equations.

Scale-free network topologies

The confluent interest of several scientific disciplines in the many aspects of the problem of Structure-Function correlations in systems made up of discretely many nonlinearly interacting components (of which gene regulatory networks are but a particular example), reccomends to pay some attention to general abstract models. These models should be both, conceptually simple and universal in their perceptions.

The universality of both saturability of the interactions and scale-free character of the interconnections among constituents in many real world systems, irrespective of the diverse nature of their components, interactions and time scales, motivated the study undertaken in [44], aimed at capturing in a simple model some general ingredients of the entangled complexity which arises from the competition of nonlinear interactions on top of complex connectivity topologies. In the model, each constituent unit (say substrate, or agent) sit on a node i ($= 1, \ldots N$) of a "small world" and "scale-free" network, and its activity level is a real function $g_i(t)$ of time. The interaction ($i \leftarrow j$) can be either activatory or inhibitory, and correspondingly the entry W_{ij} of the interaction matrix is defined respectively as $+1$ or -1, whenever interaction exists (or zero otherwise). The fraction of inhibitory interactions is a tunable parameter p of the model. The equations of motion for the activity vector $\mathbf{G}(t) = (g_1(t), \ldots g_N(t))$ are given by

$$\frac{d\mathbf{G}(t)}{dt} = -\mathbf{G}(t) + \alpha \mathbf{F}(\mathbf{W}\mathbf{G}(t)), \tag{7}$$

where $\mathbf{F}$ is a saturated response (Michaelis-Menten) vectorial function of the product of the interaction matrix $\mathbf{W}$ and the activity vector $\mathbf{G}$, and α (positive) is the interaction strength. The model has been extensively analyzed using numerical procedures which combine powerful methods of the nonlinear dynamics of many degrees of freedom and the statistical characterizations of Complex Networks. Note that for each given set of model parameters, one has to sample in network realizations and initial conditions, $\mathbf{G}(t = 0)$, in the N-dimensional phase space. Among the wealth of results obtained, we highlight here the following generic features:

- *Fluctuations over Stasis:* For intermediate values of the parameter p, the measure in phase space of asymptotic trajectories in which activities fluctuate (either periodically or chaotically) in time, prevails over stationary states.
- *Fragmentation:* In the asymptotic states (attractors) the network is fragmented, *i.e.* partitioned into islands of non-zero activity. Islands denote subgraphs which are interconnected between them by nodes which have evolved to null activity, so the dynamics of the islands are effectively disconnected. The genericity of the phenomenon of fragmentation extend to the tangent space: the instabilities experienced by the attractors when model parameters are varied, induce a partition of the dynamical island into (a) participating nodes, and (b) indifferent nodes. The later are nodes such that small single-node perturbations localised on them are orthogonal to instability.
- *Clustering:* The islands of positive activity invariably have a high clustering coefficient, which saturates with increasing network size N, in contrast to the underlying original network whose clustering coefficient tends to zero values when N is increased. Moreover, the islands inherit from the original network the small-world and scale-free character, but it is the development of dynamical fluctuations what generates the clusterization of nodes.

CONCLUSIONS

In summary, a large number of systems have been studied in the last several years from the network perspective [5]. This approach has allowed the understanding of the effects of complex topologies in many well-studied problems. By comparing the results obtained with other topologies and those for real graphs in processes such as the spreading of epidemic diseases [16] and the tolerance of complex networks to random failures and attacks, we have realized that topology plays a fundamental role.

As we have argued throughout the present overview, biological networks are not an exception. They are, at all levels of organization, the subject of intense experimental and theoretical research. The topological analyses performed on these networks have provided new useful insights [13] and are expected to produce new tools to solve longstanding problems. For instance, it is believed that a better comprehension of gene and protein networks will help to elucidate the functions of a large fraction of proteins whose functions are unknown up to date. Moreover, it is a major challenge the discovery of how biological entities interact to perform specific biological processes and tasks, as well as how their functioning is so robust under variations of internal and external parameters. Such an achievement is only possible by merging the new knowledge gained from network analysis with nonlinear dynamics models relevant in biological processes. This is what is driving current theoretical efforts, in which new mathematical models and methods borrowed from nonlinear dynamics are being studied on top of the real architecture of biological networks.

ACKNOWLEDGMENTS

J.G.G. and Y.M. are supported by MEC through a FPU grant and the Ramón y Cajal Program, respectively. This work has been partially supported by the Spanish DGICYT Projects FIS2004-05073-C04-01, FIS2005-00337 and by DGA through a research grant to FENOL.

REFERENCES

1. L. H. Harwell, J. J. Hopfield, S. Leibler and A. W. Murray, Nature **402**, C47 (1999).
2. D. B. Searls, Nature Reviews **4**, 45 (2005).
3. B. Vogelstein, D. Lane, and A. J. Levine, Nature **408**, 307 (2000).
4. H. Jeong, B. Tombor, R. Albert, Z.N. Oltvai, A.-L. Barabási, Nature **407**, 651 (2000).
5. S. Boccaletti, V. Latora, Y. Moreno, M. Chavez, and D.-U. Hwang, *Complex networks: Structure and dynamics*, Phys. Rep. **424**, 175 (2006).
6. S.A. Wagner and D.A. Fell, Proc. R. Soc. London **B268**, 1803 (2001).
7. H. Ma and A.-P. Zeng, Bioinformatics **19**, 270 (2003).
8. H.-W. Ma and A.-P. Zeng, Bioinformatics **19**, 1423 (2003).
9. M. Arita, Proc. Natl. Acad. Sci. USA **101**, 1543 (2004).
10. R. Tanaka, Phys. Rev. Lett. **94**, 168101 (2005).
11. P. Uetz, L. Glot, G. Cagney, T. A. Mansfield and E. Al., Nature **403**, 623 (2000).
12. T. Ito, T. Chiba, R. Ozawa, M. Yoshida, M. Hattori and Y. Sakaki, Proc. Natl. Acad. Sci. USA **98**, 4569 (2001).
13. H. Jeong , S.P. Mason, A.-L. Barabási, Z.N. Oltvai, Nature **411**, 41 (2001).
14. J. M. Stuart *et al.*, Science **302**, 249 (2003).
15. H. Agrawal, Phys. Rev. Lett. **89**, 268702 (2002).
16. Y. Moreno, R. Pastor-Satorras, and A. Vespignani, Eur. Phys. J. **B26**, 521 (2002).
17. F. Jacob and J. Monod, J. Mol. Bio. **3**, 318 (1961).
18. M. E. Wall, W. S. Hiavacek and M. A. Savageau, Nature Review Genetics **5**, 34 (2004).
19. S.A. Kauffman, J. Theor. Biol. **22**, 437 (1969).
20. U. Bastolla and G. Parisi, Physica D **15**, 203 (1998).
21. U. Bastolla and G. Parisi, Physica D **15**, 219 (1998).
22. J.E.S. Socolar and S.A. Kauffman, Phys. Rev. Lett. **90**, 68702 (2003).
23. B. Samuelsson and C. Troein, Phys. Rev. Lett. **90**, 98701 (2003).
24. B. Drossel, T. Mihaljev and F. Greil, Phys. Rev. Lett. **94**, 88701 (2005).
25. B. Drossel, Phys. Rev. E **72**, 16110 (2005).
26. R. ALbert and H.G. Oltmer, J. Theor. Biol. **223**, 1 (2003).
27. S. Kauffman, C. Peterson, B. Samuelsson and C. Troein, Proc. Natl. Acad. Sci. U.S.A. **100**, 14796 (2003).

28. S. Bornholdt, Science **310**, 449 (2005).
29. B. Diu, C. Guthmann, D. Lederer, and B. Roulet, *Physique Statistique* (Hermann, Paris, 1989); I. Langmuir, J. Chem. Soc. **40**, 1361 (1918).
30. B. Drossel and A.J. McKane, in *Handbook of Graphs and Networks: From the Genome to Internet*, edited by S. Bornholdt and H.G. Schuster (Wiley-VCH, Berlin, 2002); C.S. Holling, Mem. Ent. Soc. Can. **45**, 1 (1965).
31. J. Hofbauer and K. Sigmund, *Evolutionary Games and Population Dynamics* (Cambridge University Press, Cambridge, 1998).
32. N. Boccara, *Modeling Complex Systems* (Springer, New York, 2004).
33. L.A. Segel, *Modeling Dynamic Phenomena in Molecular and Cellular Biology* (Cambridge University Press, Cambridge, 1984); L. Michaelis and L.M. Menten, Biochem. Z. **49**, 333 (1913).
34. J.P. Sethna, *Statistical Mechanics: Entropy, Order Parameters, and Complexity* (Oxford University Press, New York, 2006).
35. M.B. Elowitz and S. Leibler, Nature **403**, 335 (2000).
36. J. Garcia-Ojalvo, M.B. Elowitz, and S.H. Strogatz, Proc. Natl. Acad. Sci. USA **101**, 10955 (2004).
37. K. Klemm and S. Bornholdt, Proc. Natl. Acad. Sci. USA **102**, 18414 (2005).
38. O. Bradman, J.E. Ferrel Jr., R. Li, T. Meyer, Science **310**, 496 (2005).
39. A. Ma'ayan et al., Science **309**, 1078 (2005).
40. G. von Dassow et al., Nature **406**, 188 (2000).
41. E. Mjolsness, D.H. Sharp, J. Reitnitz, J. Theor. Biol. **152**, 429 (1991; J. Reitnitz, E Mjolsness, D.H. Sharp, J. Exp. Zoology **271**, 47 (1992).
42. V.V. Gursky, J. Reitnitz, A.M. Samsonov, Chaos **11**, 132 (2001).
43. R.V. Solé, I. Salazar-Ciudad, J. García-Fernández, Physica A **305**, 640 (2002); I. Salazar, J. García-Fernández, R.V. Solé, J. Theor. Biol. **205**, 587 (2000); R.V. Solé and R. Pastor-Satorras, in *Handbook of Graphs and Networks: From the Genome to Internet*, edited by S. Bornholdt and H.G. Schuster (Wiley-VCH, Berlin, 2002).
44. J. Gómez-Gardenes, Y. Moreno, L.M. Floría, Chaos **16**, 015114(1-11) (2006); Biophys. Chem. *115*, 225 (2005); Physica A **352**, 265 (2005).

Unspecific Cooperative Ligand Binding to One-Dimensional Lattice-like Macromolecules

Adrian Velazquez-Campoy

Institute of Biocomputation and Complex Systems Physics (BIFI)
Universidad de Zaragoza, Corona de Aragón 42, E-50009 Zaragoza, Spain
Phone: +34 976 562215
Fax: +34 976 562215
e-mail: adrianvc@unizar.es

Abstract. Unspecific ligand binding to one-dimensional lattice-like macromolecules, a common event in nature (e.g. ligand/protein binding to DNA and certain carbohydrates), presents distinctive features when compared to specific ligand binding to macromolecules. McGhee and von Hippel developed a mathematical formalism in the form of the Scatchard representation. This article presents the application of the theory for unspecific cooperative ligand binding to linear lattice-like macromolecules in isothermal titration calorimetry following an exact and accurate method, without the limitations and deficiencies of the Scatchard formalism.

Keywords: cooperative ligand binding, lattice-like macromolecule, isothermal titration calorimetry
PACS: 87.14.Ee, 87.14.Gg, 87.15.Kg

INTRODUCTION

Many macromolecular interactions involve the non-specific binding of ligands onto a one-dimensional lattice-like macromolecule (e.g., nucleic acid or carbohydrate ligand binding). The linear macromolecule is considered as a mono-dimensional lattice constituted of N repeating units, and any bound ligand molecule occupies a certain number, l, of these repeating units (see Figure 1). Thus, the potential binding sites overlap and are homogeneously distributed throughout the macromolecule. The geometric properties of the system, that is, N, the number of repeating units per macromolecule conforming each binding site, and l, the length of the ligand (in a macromolecule repeating unit basis), besides the thermodynamic binding parameters, govern the behavior of the system.

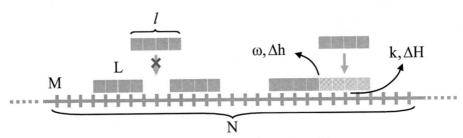

CP851, *From Physics to Biology; BIFI 2006 II International Congress,*
edited by J. Clemente-Gallardo, Y. Moreno, J. F. Sáenz Lorenzo, and A. Velázquez-Campoy
© 2006 American Institute of Physics 0-7354-0350-3/06/$23.00

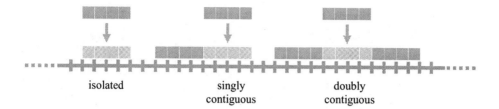

isolated   singly   doubly

contiguous   contiguous

FIGURE 1. Schematic view of ligand binding to a one-dimensional lattice-like macromolecule. The macromolecule is composed of N repeated units and any bound ligand occupies l units. If the spacing between to bound ligands is smaller than l units, then no ligand can bind between them. If a ligand binds adjacently to an already bound ligand, then the energetics of the binding reflects two contributions: the interaction ligand-macromolecule (dissociation constant k and binding enthalpy ΔH) and the interaction ligand-ligand (cooperativity interaction constant ω and cooperativity interaction enthalpy Δh). There are three binding modes: isolated ligand with no interacting neighbors, singly contiguous ligand with one interacting neighbor, and doubly contiguous ligand with two interacting neighbors. Other possibilities (e.g. directional binding for asymmetric ligands or partial ligand binding are not considered).

If a ligand binds to the macromolecule, its interaction can be described in terms of the binding affinity or dissociation constant, k, and the binding enthalpy, ΔH. If the spacing between two bound ligands is smaller than the ligand size, l, then, no ligand may bind between them. If there is no ligand-ligand interaction when ligands are bound, no additional thermodynamic parameters are needed to describe the system. Otherwise, two additional thermodynamic parameters are required: the cooperativity interaction constant ω and cooperativity interaction enthalpy Δh, which describe the interaction between adjacent bound ligands.

Ligand binding is usually described using the number of ligand molecules bound per macromolecule, ν, defined as:

$$\nu = \frac{[L]_B}{[M]_T} \tag{1}$$

where $[L]_B$ is the concentration of bound ligand and $[M]_T$ is the total concentration of macromolecule. The limiting values for the binding parameter are 0 and N/l, thus, reflecting the degree of saturation of the macromolecule. The difficulty in studying this special type of ligand binding stems from the fact that the intuitive interpretation in ligand binding to a macromolecule with l equivalent and independent binding sites does not apply here: the number of free binding sites is not equal to the maximum number of binding sites $(N - l + 1)$ minus the number of occupied binding sites (ν), because binding sites overlap and bound ligands may leave free repeating units in the macromolecule which cannot be occupied. Figure 2 illustrates the effect of overlapping binding sites and the size of the ligand on the binding of a ligand to a linear macromolecule. For ligand sizes larger than 1, an *apparent* negative cooperativity effect appears: saturation slows down at high free ligand concentration and the Scatchard plot is not linear.

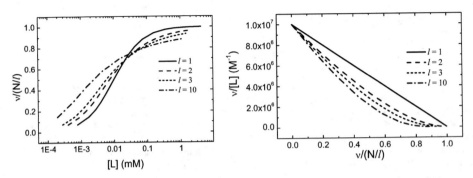

FIGURE 2. Simulations corresponding to a macromolecule with N = 100 binding sites, dissociation constant k = 10^{-5} M, and ligand with size l = 1,2,3,10. The number of ligand molecules bound, v, has been normalized by the factor N/l in order to compare better between the four cases.
(left panel) Evolution of the number of ligand molecules bound to the macromolecule as the free ligand concentration increases. If the ligand size l is large, the number of ligand molecules bound increases fast at low saturation (bound ligands occupy larger stretches), but it slows down at high saturation, suggesting an apparent negative cooperativity. This is usually termed "entropic resistance" and it can be explained on a statistical basis: the larger the ligand size, the more difficult to find empty stretches large enough to bind.
(right panel) Scatchard representation of the same data shown in the left panel. The apparent negative binding cooperativity (indicated by a positive curvature and the separation from the linear plot) is observed as the ligand size l increases.

McGhee and von Hippel derived a closed form, in the Scatchard representation, valid for any ligand size, l, any ligand dissociation constant, k, and any level of cooperativity interaction between ligands bound contiguously, ω, for infinite homogeneous lattices [1,2]:

$$\frac{v}{[L]} = f(v; N, l, k, \omega) \tag{2}$$

Extensions of this theory have been developed for finite lattices [3,4], heterogeneous systems in which the binding sites are not homogeneously distributed throughout the macromolecule [5], different classes of binding sites are present in the macromolecule [4], different ligand binding modes [3,6,7], a mixture of different ligands, which could be used in displacement experiments [8], or an allosteric binding model [9,10]. This theory has been barely applied in isothermal titration calorimetry, and inappropriately applied in many cases. Moreover, cooperative binding has never been implemented.

Isothermal titration calorimetry (ITC) is unique among the titration techniques, because it allows the simultaneous determination of the affinity, k, and the enthalpy of binding, ΔH. Therefore, it is possible to perform a complete characterization of the binding process (determination of affinity, Gibbs energy, enthalpy and entropy of binding) in just one experiment. Both spectroscopy and calorimetry allow evaluating the binding affinity. In addition, calorimetry measures the heat of reaction at constant pressure, that is, it provides the binding enthalpy. This thermodynamic potential is important in describing the intermolecular driving interactions underlying the binding process and it can be of help when discriminating experimentally between different binding models. Despite the widespread use of ITC for studying biomolecular reactions, it has been scarcely applied to lattice-like macromolecular systems and its full potential has not been completely exploited [4,11-19] and very few times used properly [4,13-15,17,19]. Two situations will be considered: independent binding and cooperative binding.

INDEPENDENT BINDING

In the case of a homogeneous mono-dimensional lattice-like macromolecule with N independent binding sites and ligand with size l ($l \geq 1$), the following expression holds:

$$\frac{v}{[L]} = \frac{N - lv}{k} \left(\frac{N - lv}{N - (l-1)v} \right)^{l-1} \tag{3}$$

This equation was derived assuming an infinite lattice; however, real macromolecules are not infinite and end-effects will be present. However, the error is estimated to be less than the experimental error if $N/l > 30$. Introducing in Eq. 3 a correction factor $(N-l+1)/N$ valid for finite lattices, or employing an exact combinatorial analysis for finite lattices [3], it can be demonstrated that the infinite lattice approximation is good even if N/l is significantly lower than 30. The previous equation can be rewritten as an ($l+1$)-order polynomial equation in v:

$$P(v) = ([L]_T - [M]_T v)(N - lv)^l - kv(N - (l-1)v)^{l-1} = 0 \tag{4}$$

For any given values for $[L]_T$, $[M]_T$ and k, a root with physical meaning (i.e. in the $[0,N/l]$ interval) can be determined employing numerical methods (e.g., Newton-Raphson).

The heat effect associated with injection, q_i, can be evaluated as follows:

$$q_i = V\Delta H \left([M]_{T,i} v_i - \left(1 - \frac{v}{V} \right) [M]_{T,i-1} v_{i-1} \right) \tag{5}$$

where V is the calorimetric cell volume and v is the injection volume. In order to calculate q_i values and compare them with the experimental values, it is necessary to calculate the binding parameter after each injection, v_i, knowing $[M]_{T,i}$ and $[L]_{T,i}$, and assuming values for N, l and k. Thus, Eq. 5 is used in a non-linear regression procedure to extract the optimal values of the thermodynamic parameters N, l, k and ΔH from the experimental data.

165

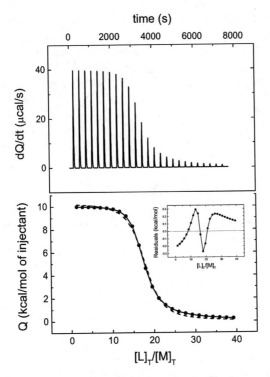

FIGURE 3. Simulated calorimetric titration corresponding to a macromolecule with N = 100 binding sites and a ligand with size $l = 5$. The calculations were done with a concentration of ligand in syringe $[L]_0 = 10$ mM, a concentration of macromolecule in the calorimetric cell $[M]_0 = 0.05$ mM, a dissociation constant of k = 10^{-5} M and a binding enthalpy $\Delta H = 10$ kcal/mol. In all calculations, an injection volume v = 10 μL and a calorimetric cell volume V = 1.4 mL have been used. In the thermogram, the upper plot, the signal directly monitored in the experiment, that is, the thermal power applied by the feedback system in order to maintain a constant temperature irrespective of what is occurring in the cell, is shown. Each peak in the sequence corresponds to one injection of ligand into the macromolecule solution. In the lower plot the integrated area of each peak is represented as a function of the molar ratio, i.e. the ratio between the total concentration of ligand and the total concentration of macromolecule in the calorimetric cell after any injection. The thermodynamic parameters, N, l, k, and ΔH, would be determined by performing a non-linear regression analysis of the titration plot. (Inset) Residual plot after data fitting with a model for a ligand with size $l = 1$.

Figure 3 shows the simulated titration corresponding to a macromolecule with N = 100 binding sites and a ligand with size $l = 5$. There exist some significant differences between this titration and the one corresponding to a system with ligand size $l = 1$. The expected maximal number of ligand molecules bound per macromolecule is N/l = 100/5 = 20. However, the apparent maximal binding parameter, inferred from the localization of the inflection point of the titration plot, is around 17. The larger the ligand size, the larger the difference between the apparent and the actual maximal binding parameters. Therefore, if the data analysis is performed with a model for a ligand with size $l = 1$ [30,32], then, both the maximal binding parameter and the binding affinity will be estimated with a significant error: N = 17.2, k = 1.2·10^{-5} M and ΔH = 10.4 kcal/mol; whereas the values used for that calculation where: N = 100, l = 5, N/l = 20, k = 10^{-5} M and ΔH = 10 kcal/mol. Moreover, the theoretical curve corresponding to ligand size $l = 1$ differs significantly from the curve for ligand $l = 5$, as seen with the residuals plot (see inset in Figure 3). The larger the ligand size l, the larger the discrepancy.

COOPERATIVE BINDING

If bound ligands interact, favorably or unfavorably, with nearest neighbors on either side, an interaction or cooperativity parameter, ω, defined as an equilibrium constant between two states (two ligands interacting side-by-side and the same two ligands not interacting and separated at least by one free macromolecule unit) is included representing such interaction energy. Again, a close analytical expression for the binding parameter has been derived [1,2]:

$$\frac{v}{[L]} = \frac{N - lv}{k}\left(\frac{(2\omega - 1)(N - lv) + v - R}{2(\omega - 1)(N - lv)}\right)^{l-1}\left(\frac{N - (l+1)v + R}{2(N - lv)}\right)^2 \tag{6}$$

with:

$$R = \left((N - (l+1)v)^2 + 4\omega v(N - lv)\right)^{1/2} \tag{7}$$

The previous equation can be rewritten as an (l+3)-order polynomial equation in v:

$$P(v) = ([L]_T - [M]_T v)(N - lv)((2\omega - 1)(N - lv) + v - R)^{l-1}(N - (l+1)v + R)^2$$
$$- kv(2(\omega - 1)(N - lv))^{l-1}(2(N - lv))^2 = 0 \tag{8}$$

For any given values for $[L]_T$, $[M]_T$ and k, a root with physical meaning (i.e. in the [0,N/l] interval) can be determined employing numerical methods (e.g., Newton-Raphson).

The heat effect associated with injection, q_i, can be evaluated as follows:

$$q_i = V\left(\Delta H\left([M]_{T,i} v_{isol,i} - \left(1 - \frac{v}{V}\right)[M]_{T,i-1} v_{isol,i-1}\right)\right.$$
$$+ \left(\Delta H + \frac{\Delta h}{2}\right)\left([M]_{T,i} v_{sc,i} - \left(1 - \frac{v}{V}\right)[M]_{T,i-1} v_{sc,i-1}\right) \tag{9}$$
$$\left. + (\Delta H + \Delta h)\left([M]_{T,i} v_{dc,i} - \left(1 - \frac{v}{V}\right)[M]_{T,i-1} v_{dc,i-1}\right)\right)$$

where Δh is the enthalpy associated with the interaction between nearest neighbor bound ligands and v_{isol}, v_{sc} and v_{dc} are the partial number of ligand molecules bound isolate with no neighbors, with only one nearest neighbor (singly-contiguous) and with two nearest neighbors (doubly contiguous), per macromolecule, respectively. In order to calculate q_i values and compare them with the experimental values, it is necessary to calculate the binding parameter after each injection, v_i, knowing $[M]_{T,i}$ and $[L]_{T,i}$, and assuming values for N, l, k and ω. Thus, Eq. 9 is used in a non-linear regression procedure to extract the optimal values of the thermodynamic parameters N, l, k, ΔH, ω and Δh from the experimental data.

The partial binding parameters can be calculated, once the total binding parameter has been determined, as follows:

$$
\begin{aligned}
v_{isol} &= \left([L]_T - [M]_T\, v\right) \frac{N - lv}{k} \left(\frac{(2\omega - 1)(N - lv) + v - R}{2(\omega - 1)(N - lv)} \right)^{l+1} \\[2mm]
v_{sc} &= \left([L]_T - [M]_T\, v\right) \frac{\omega}{\omega - 1} \frac{(l-1)v - N + R}{k} \left(\frac{(2\omega - 1)(N - lv) + v - R}{2(\omega - 1)(N - lv)} \right)^{l} \\[2mm]
v_{dc} &= \left([L]_T - [M]_T\, v\right) \left(\frac{\omega}{2(\omega - 1)} \right)^2 \frac{((l-1)v - N + R)^2}{k(N - lv)} \left(\frac{(2\omega - 1)(N - lv) + v - R}{2(\omega - 1)(N - lv)} \right)^{l-1}
\end{aligned}
\tag{10}
$$

and the sum of these partial binding parameters must equal the total number of ligand molecules bound to the lattice:

$$
v = v_{isol} + v_{sc} + v_{dc}
\tag{11}
$$

The interaction enthalpy is the enthalpy associated to the interaction between two adjacent bound ligands. Because bound ligands are indistinguishable, in order to prevent counting twice such interaction when evaluating the heat associated with binding using Eq. 9, the contribution of Δh is divided by 2 in the terms corresponding to singly and doubly contiguous bound ligands.

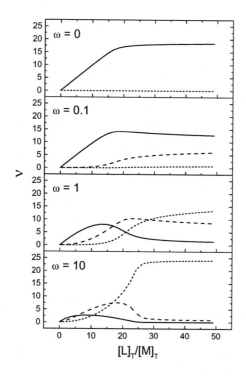

FIGURE 4. Effect of the cooperativity parameter on the evolution of the partial binding parameters along the titration. The calculations were done with a number of binding sites per macromolecule N = 100, a ligand size l = 4, a concentration of ligand in syringe $[L]_0$ = 25 mM, a concentration of macromolecule in the calorimetric cell $[M]_0$ = 0.1 mM, a dissociation constant of k = 10^{-5} M and a cooperativity parameter ω = 0, 0.1, 1 and 10. For each value of the cooperativity parameter ω, the partial binding parameters are calculated: v_{isol} (continuous line), v_{sc} (dashed line) and v_{dc} (dotted line).

Figure 4 illustrates the effect of the cooperativity parameter on the evolution of the partial binding parameters along the titration. The calculations were done with a number of binding sites per macromolecule N = 100 and a ligand size l = 4. Four values of the cooperativity parameter have been considered: ω = 0 (maximal negative cooperativity), 0.1 (negative cooperativity), 1 (no cooperativity) and 10 (positive cooperativity). In the case with no cooperativity, isolatedly bound ligands appear from the beginning, then bound ligands with only one neighbor (singly contiguous) and later bound ligands with two neighbors (doubly contiguous). When ligands bind with positive cooperativity, isolatedly bound ligands appear first, but the scene is dominated soon by ligands bound with two neighbors. When ligands bind with negative cooperativity, isolatedly bound ligands dominate all over the titration. When ligands bind with maximal negative cooperativity there are no ligands adjacently bound at all (this case is completely equivalent to ligand size l_{app} = l+1 and no binding cooperativity).

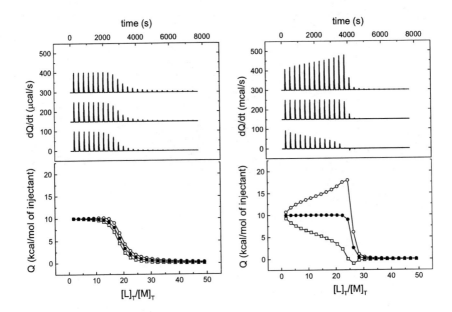

FIGURE 5. Effect of the interaction enthalpy on the calorimetric titration. Simulated calorimetric titrations corresponding to a macromolecule with N = 100 binding sites and a ligand size l = 4. The calculations were done with a concentration of ligand in syringe $[L]_0$ = 25 mM, a concentration of macromolecule in the calorimetric cell $[M]_0$ = 0.1 mM, a dissociation constant of k = 10^{-5} M, a binding enthalpy ΔH = 10 kcal/mol, a cooperativity parameter ω = 0.1 and 10, and an interaction enthalpy Δh = -5 (open squares), 0 (closed circles) and 5 kcal/mol (open circles).

(left panel) Calorimetric titrations with ω = 0.1 (negative cooperativity). The effect of the interaction enthalpy is small, because ligands tend to bind isolatedly; hence the heat effect reflects the binding almost exclusively.

(right panel) Calorimetric titrations with ω = 10 (positive cooperativity). The effect of the interaction enthalpy is remarkable, because ligands tend to bind clustered; hence the heat effect reflects both the binding event plus the interaction between nearest-neighbor bound ligands.

In Figure 5 the effect of a non-zero interaction enthalpy is illustrated. The interaction enthalpy is determinant in discriminating between different cases and it is one of the advantages of using ITC to study binding reactions. If Δh is equal to zero, there are two differences between negative and positive cooperativity: unfavorable ligand-ligand interaction lowers both the apparent affinity (steepness of the titration plot) and the apparent stoichiometry (localization of the equivalency or inflection point). Thus, bearing in mind that the apparent stoichiometry is also modulated by the actual N and l values, without any *a priori* information there would be a problem to decide which case applies to the system under study (e.g. low affinity with no cooperativity would be equivalent to high affinity with negative cooperativity; small ligand size with no cooperativity would be equivalent to large ligand size with positive cooperativity). At this point, it must be remembered that the parameter l is usually considered as an effective ligand size, incorporating implicitly ligand-ligand interaction effects. However, the presence of Δh (which can be modified changing the

170

experimental conditions) will help in selecting the appropriate case. If Δh is non-zero, there is an additional contribution to the heat effect coming from the interaction between adjacently bound ligands and the different cases can be distinguished. Even if, under certain circumstances, the interaction enthalpy is zero, the enthalpy associated to any process is in general very sensitive to the environmental experimental conditions (temperature, pH, ionic strength, etc.) and, therefore, easily modulated attaining a non-zero value. If there is negative cooperativity the effect of non-zero enthalpy is small, because the population of ligands bound with nearest neighbors increase very slowly as the titration proceeds (see Figure 5). On the other hand, if there is positive cooperativity the effect of non-zero enthalpy is dramatic (a marked increase or decrease of the heat effect in the pre-equivalency region for positive or negative interaction enthalpy, respectively), because the population of ligands bound with at least one nearest neighbor dominates the titration (see Figure 5).

As expected, the consequences of ligand binding cooperativity are more dramatic at high interaction constant values. If the interaction constant is moderate ($1 < \omega < 20$), at the beginning of the titration any ligand tends to bind isolate and this is reflected in that the heat detected corresponds mainly to the binding enthalpy ΔH; then, in the intermediate region the population of bound ligands with one neighbor increases and the heat detected corresponds to the binding enthalpy plus the interaction cooperativity enthalpy $\Delta H + \Delta h$; later on, as saturation is reached, the population of bound ligands with two neighbors increases and the heat detected corresponds to the binding enthalpy plus twice the interaction cooperativity enthalpy, $\Delta H + 2\Delta h$ (Figure 6). On the other hand, if the interaction constant is large ($\omega > 100$), from the beginning of the titration ligands tend to bind clustered so that on average almost any ligand has two neighbors (except the ligands at both ends of the clusters) and this is reflected in that the heat detected corresponds to the binding enthalpy plus the interaction cooperativity enthalpy, $\Delta H + \Delta h$ (not twice the cooperativity enthalpy, since in that case interactions ligand-ligand would be counted twice). Therefore, in the limit of extremely high cooperativity interaction constant ($\omega > 1000$), the titration would be equivalent to that of a ligand with an extremely high affinity and a binding enthalpy equal to $\Delta H + \Delta h$.

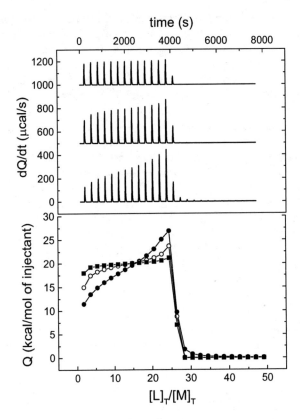

FIGURE 6. Effect of high interaction cooperativity constant values on the calorimetric titration. Simulated calorimetric titrations corresponding to a macromolecule with N = 100 binding sites and a ligand size l = 4. The calculations were done with a concentration of ligand in syringe $[L]_0$ = 25 mM, a concentration of macromolecule in the calorimetric cell $[M]_0$ = 0.1 mM, a dissociation constant of k = 10^{-5} M, a binding enthalpy ΔH = 10 kcal/mol, an interaction enthalpy Δh = 10 and a cooperativity parameter ω = 10 (closed circles), 100 (open circles) and 1000 (closed squares).

REVERSE TITRATION *vs.* DIRECT TITRATION

Reverse titrations may be used to provide additional insight on the ligand-macromolecule interaction and confirm results obtained through direct titrations in ligand binding in macromolecules with different classes of binding sites or interacting binding sites. Therefore, more information on the interaction or cooperativity thermodynamic parameters can be obtained by performing reverse titrations [4,17,20]. For example, it is possible to discriminate between specific and unspecific ligand binding to DNA occurring simultaneously [17]. In a direct titration the ligand is injected into a macromolecule solution, whereas in a reverse titration the macromolecule is injected into a ligand solution. If the stoichiometry is 1:1, no differences should arise between these two experiments. However, if the stoichiometry is not 1:1 and the binding sites are not equivalent and/or independent, the comparison of both experiments reveals substantial differences.

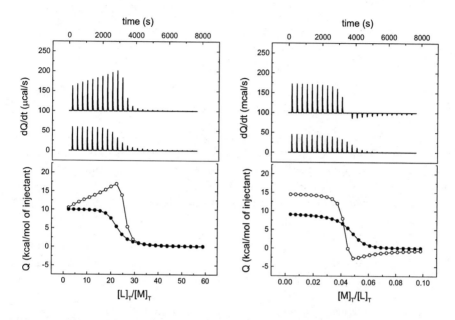

FIGURE 7. Comparison between direct titrations (ligand into macromolecule solution; left panel) and reverse titrations (macromolecule into ligand solution; right panel). Simulated calorimetric titrations corresponding to a macromolecule with N = 100 binding sites and a ligand size l = 4. For the direct titrations, a concentration of ligand in syringe $[L]_0$ = 15 mM and a concentration of macromolecule in the calorimetric cell $[M]_0$ = 0.05 mM were employed. For the reverse titrations, a concentration of ligand in the calorimetric cell $[L]_0$ = 1 mM and a concentration of macromolecule in the syringe $[M]_0$ = 0.5 mM were employed. A dissociation constant of k = 10^{-5} M, a binding enthalpy ΔH = 10 kcal/mol, a cooperativity parameter ω = 1 (closed circles) and 5 (open circles) (negative cooperativity has not been simulated, because of its small effect, as demonstrated in Figure 5), and an interaction enthalpy Δh = 0 and 5 kcal/mol were used. In the reverse titrations the heat associated with each injection was normalized per mole of binding sites injected, that is, dividing the heat per injection by v × $[M]_0$ × N/l.

Figure 7 shows direct (left panel) and reverse (right panel) calorimetric titrations with and without ligand cooperativity. In the initial stages of a direct titration the macromolecule is in excess and the ligand is at a subsaturating concentration. The evolution of the free and bound species (ligand and macromolecule) is markedly different in a reverse titration compared to a direct titration. As the titration proceeds, the ligand reaches higher concentrations and the macromolecule becomes saturated, the number of ligand molecules bound per macromolecule increases, the number of ligands isolatedly bound per macromolecule decreases, and the number of ligands doubly contiguous increases. On the other hand, in the initial stages of a reverse titration the ligand is in excess and the macromolecule becomes almost saturated at the beginning, and, therefore, most of the ligands bound present nearest neighbors (singly and doubly contiguous). As the titration proceeds, the macromolecule reaches higher concentrations and the ligand subsaturating concentrations. Consequently, the increase in macromolecule concentration induces ligand dissociation. Then, some of the bound ligands with nearest neighbors get dissociated to bind with fewer neighbors. Therefore, the number of ligand molecules bound per macromolecule decreases, the number of ligands isolatedly bound increases, and the number of ligands doubly contiguous decreases. Therefore, if there is cooperativity, there are important differences between both experiments.

CONCLUSIONS

McGhee - von Hippel theory for cooperative ligand binding to one-dimensional homogeneous lattice-like macromolecules has been implemented in isothermal titration calorimetry. The method is exact and rigorous, without the weaknesses and drawbacks intrinsic to the Scatchard analysis, and it can be employed in any titration technique (e.g. spectroscopy) in which the only independent variables are the total concentration of reactants. The procedure requires solving numerically an $(l+3)$-order or an $(l+1)$-order polynomial equation (where l is the ligand size), depending on whether or not cooperative binding is considered. Extensions for heterogeneous lattice-like macromolecules (e.g. finite lattices, non-homogeneous distribution of binding sites or different classes of binding sites in the macromolecule) can easily be done.

This methodology permits the study of binding reactions with long one-dimensional macromolecules (e.g. DNA) by ITC, which has been scarcely used for this type of systems, thus, benefiting from using this technique. ITC allows determining simultaneously the affinity and the enthalpy of binding, discriminating between intrinsic parameters (k and ΔH) and interaction or cooperativity parameters (ω and Δh), and, therefore, it provides a complete thermodynamic characterization of the binding process (Gibbs energy, enthalpy, entropy and stoichiometry). Besides, the possibility of modulating affinity and enthalpy by changing the experimental conditions, gives us an additional element to extract valuable information and to discriminate among different possible cases applicable to the system under study. The comparison between reverse and direct titrations gives additional information about cooperativity phenomena. On the other hand, limitations intrinsic to ITC should be bore in mind (e.g. kinetic effects arising in slow binding reactions).

As it has been pointed out, the implemented model is rather simple and does not incorporate some features that may be important in some cases. For example, asymmetric ligands binding directionally onto the macromolecule will give rise to different interaction parameters depending on the orientation of a given ligand binding relative to its potential neighboring ligands. Another aspect to consider is the possibility for a ligand to bind partially if the distance between to bound ligands is smaller than l macromolecule units. Therefore, there is enough room for further refinements of the methodology presented here.

ACKNOWLEDGMENTS

A.V.C is a recipient of a Ramón y Cajal Research Contract from the Spanish Ministry of Science and Technology. This research was funded by grant SAF2004-07722 from the Spanish Ministry of Education and Science.

REFERENCES

1. J.D. McGhee, and P.H. von Hippel. *J. Mol. Biol.* **86**, 469-489 (1974).

2. J.D. McGhee, and P.H. von Hippel. *J. Mol. Biol.* **103**, 679 (1976).

3. I.R. Epstein. *Biophys. Chem.* **8**, 327-339 (1978).

4. O.V. Tsodikov, J.A. Holbrook, I.A. Shkel, and M.T. Record Jr. *Biophys. J.* **81**, 1960-1969 (2001).

5. A. Kristiansen, K.M. Vårum, and H. Grasdalen. *Biochim. Biophys. Acta.* **1425**, 137-150 (2001).

6. G. Schwarz, and S. Stankowski. *Biophys. Chem.* **10**, 173-181 (1979).

7. S. Rajendran, M.J. Jezewska, W. Bujalowski. *J. Biol. Chem.* **273**, 31021-31031 (1998).

8. C.R. Cantor, and P.R. Schimmel. *In Biophysical chemistry, Part III: The behaviour of biological macromolecules.* W.H. Freeman, New York, U.S.A. (1980).

9. N. Dattagupta, M. Hogan, and D.M. Crothers. *Biochemistry.* **19**, 5998-6005 (1980).

10. J.B. Chaires. *J. Biol. Chem.* 261:8899-8907 (1986).

11. T. Ohyama, and J.A. Cowan. *J. Biol. Inorg. Chem.* **1**, 83-89 (1996).

12. T. Ohyama, and J.A. Cowan. *J. Biol. Inorg. Chem.* **1**, 111-116 (1996).

13. T. Lundbäck, and T. Härd. *J. Phys. Chem.* **100**, 17690-17695 (1996).

14. T. Lundbäck, H. Hansson, S. Knapp, R. Ladenstein, and T. Härd. *J. Mol. Biol.* **276**, 775-786 (1998).

15. M.M. Lopez, K. Yutani, and G.I. Makhatadze. *J. Biol. Chem.* **274**, 33601-33608 (1999).

16. K. Utsuno, Y. Maeda, and M. Tsuboi. et al. *Chem Pharm. Bull.* **47**, 1363-1368 (1999).

17. J.A. Holbrook, O.V. Tsodikov, R.M. Saecker, and M.T. Record Jr. *J. Mol. Biol.* **310**, 379-401 (2001).

18. M. Girard, S.L. Turgeon, and S.F. Gauthier. *J. Agric. Food Chem.* **51**, 4450-4455 (2003).

19. W.B. Peters, S.P. Edmondson, and J.W. Shriver. *J. Mol. Biol.* **343**, 339-360 (2004).

20. S.C. Kowalczykowski, L.S. Paul, N. Lonberg, J.W. Newport, J.A. McSwiggen, and P.H. von Hippel. *Biochemistry.* **25**, 1226-1240 (1986).

New Computational Approaches for NMR-based Drug Design: A Protocol for Ligand Docking to Flexible Target Sites[*]

Luis Gracia[1], Joshua A. Speidel[1] and Harel Weinstein[1,2]

[1]Department of Physiology and Biophysics, and [2]HRH Prince Alwaleed Bin Talal Bin Abdulaziz Alsaud Institute for Computational Biomedicine, Weill Medical College of Cornell University, New York, NY 10028, USA

Abstract. NMR-based drug design has met with some success in the last decade, as illustrated in numerous instances by Fesik's "ligand screening by NMR" approach. Ongoing efforts to generalize this success have led us to the development of a new paradigm in which quantitative computational approaches are being integrated with NMR derived data and biological assays. The key component of this work is the inclusion of the intrinsic dynamic quality of NMR structures in theoretical models and its use in docking. A new computational protocol *is introduced* here, designed to dock small molecule ligands to flexible proteins derived from NMR structures. The algorithm makes use of a combination of simulated annealing monte carlo simulations (SA/MC) and a mean field potential informed by the NMR data. The new protocol is illustrated in the context of an ongoing project aimed at developing new selective inhibitors for the PCAF bromodomains that interact with HIV Tat.

Keywords: ligand docking, protein flexibility, mean field potential, simulated annealing, monte carlo simulation, bromodomains, anti-HIV drugs.
PACS: 01.30.Cc, 82.56.Pp, 87.15.Aa, 02.70.Uu.

INTRODUCTION

Many computational methods predict the binding mode of small molecules in complexes with biological macromolecules using the simplifying assumption of rigid docking. It is well known, however, from ample structural data that both ligand and receptor often change conformation upon binding, a mechanism know as induced fit and proposed by Koshland in 1958 [1].

The introduction of protein flexibility in computational docking algorithms is complicated by the fluctuations of protein structures between conformational states [2]. Upon binding, a ligand might stabilize one of the states, driving the conformational equilibrium. Still, the most prevalent structural rearrangements related to the binding of ligand in the binding site have been shown to involve side chain conformations, whereas the backbone tends to undergo smaller deviations from the unbound state [3]. The introduction of these considerations in computational docking algorithms is not simple, and most recent docking algorithms do account for ligand flexibility, but protein flexibility remains a challenge.

In efforts to develop a structure-based ligand design paradigm for inhibitors of a key step in HIV-1 replication, we targeted the bromodomain (BRD) that binds the acetylated form of HIV-

---

[*] Presented in poster format at the BIFI 2006 International Conference *"From Physics to Biology: the interface between experiment and computation"*, Zaragoza, Spain.

Tat [4]. To this end, we needed to develop a method for in *silico* docking to specific BRDs. The specific target for ligand design is a member of a class of BRDs that specifically bind acetylated lysine (AcK) [5]. Structural data obtained from NMR showed that the region in the Tat protein that recognizes the P/CAF BRD in the process of HIV-1 transactivation in vivo, surrounds the acetylated lysine50 (AcK50) [6,7]. Structures of apo P/CAF BRD and its complex with a AcK50-Tat-derived peptide and with several ligands have been determined by NMR [4,7]. The P/CAF structure consists of a four helix bundle with a left-handed twist. The binding pocket for this complex has been identified to be in a hydrophobic region between two flexible loops (ZA and BC loops). The flexibility that these loops exhibit becomes evident from comparisons of their conformations in the apo and various ligand-bound forms. After superposition of the four rigid alpha helices, the RMSD of the trace for the ZA loop between the apo P/CAF and the Tat-P/CAF complex is 6.0 ± 0.6 Å for the NMR ensemble, while for the alpha helix bundle it is 1.1 ± 0.1 Å. Similarly, the RMSD for the ZA loop between the apo P/CAF and the P/CAF complexed with four different ligands is around 4 Å. Similar values of RMSD are found for the BC loop.

Clearly, ligand docking methods for this system must take into consideration the flexibility of the protein in this region. We therefore developed a method suitable for docking flexible ligands into a binding site composed of such flexible loop regions. In this brief contribution we present the method applied to four ligands designed to bind to P/CAF BRD [4].

METHODS

The docking protocol involves the generation of initially random positions and conformations of the ligand around the target active site, followed by a Monte Carlo Simulated Annealing (MC/SA) optimization. The ligand and receptor side-chains that might participate in the ligand binding process are optimized in this process, while the rest of the protein is kept fixed. In addition, the flexibility of the backbone of the loops that line the active site is accounted for by a multiple copies simulation protocol in the docking procedure.

Briefly, the multiple copy approach focuses on the interesting regions of the system by creating multiple copies of those regions. During the simulation, the atoms in the framework (i.e., outside the multiple copy regions) respond to the average force from all multiple copy regions, whereas each multiple copy region responds to the full force from the framework but feels no forces from other copies. If several multiple copy regions are considered, the copies of one region are transparent to one another, but each feels the average force exerted by copies in each of the other regions. In the present algorithm, there are two such region (the ZA and BC loops), and the multiple copies were obtained from the manifold of NMR structures. Other modes for obtaining the multiple copies can be implemented in the same manner.

To evaluate the energy of the system we employ a mean-field potential (MFP, for a review see Koehl and Delarue [8]) similar to that proposed by Bastard *et al.* [9]. Barriers separating the minima in the mean field calculation are lower than in the original system, providing a smoothing effect of the energy landscape [10], and thus improving sampling efficiency. The algorithm selects the best energy replicas for each loop, eliminating the high energy ones based on their relative probabilities, which are allowed to evolve during the MC/SA procedure. The entire method has been implemented in CHARMM [11].

We tested the method with four ligands that were designed to bind to P/CAF and for which structures are available from NMR: *N(3-aminopropyl)-2-nitrobenzenamine* (**NP2**, PDB ID: 1WUM) [4], *N(3-aminopropyl)-4-methyl-2-nitrobenzenamin* (**NP1**, PDB ID: 1WUG) [4], *N(3-*

aminopropyl)-4-metoxi-2-nitrobenzenamin **(NP3)** and *4-methyl-2H-chromene-2,7,8-triol* **(DMC)** [12]. In each case, the ligand was docked into a protein system that consisted of the alpha helix bundle from NMR model 1, and 10 copies (i.e., NMR models 1 to 10) of the loops ZA and BC. To adjust the conformations of the loops of the different NMR models on to the framework of model 1, a gentle minimization was done.

During the docking procedure, the ligand was allowed full flexibility (translation, rotation and internal torsions), and the side-chains of the multiple copy loops were allowed to move in torsional space. To measure the agreement between the computational docking and the experimental structure we calculated the RMSD of ligand heavy atom from the different corresponding NMR models.

RESULTS AND DISCUSSION

The new algorithm differs from other MC/SA flexible docking methods in that several copies of a region of the protein are considered. This complicates the pose sampling because the ligand has to find its way through the 'crowded' binding pocket created by the side-chains of the multiple copies. On the other hand, the method is more efficient than separate docking to each of the individual systems in order to select optimal poses, as the multiple copy loops are evaluated relative to each other, allowing the selection of the best candidates.

Of the four ligands we tested, three are positioned with an RMSD <2.0 Å, a threshold commonly used in docking [13] (NP2: 1.63 Å, NP3: 2.02 Å and DMC: 1.31 Å). Figure 1 shows the best docking pose for ligands NP2 and DMC as representative docking solutions.

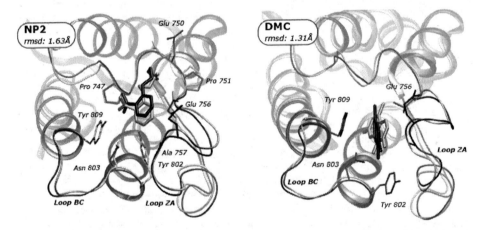

FIGURE 1. The best poses for two of the ligands NP2 (1.63Å) and DMC (1.31Å) are shown (dark gray) superimposed on the corresponding NMR model 1 (light gray). Alpha helices of the protein are shown in ribbons and the ZA and BC loops are shown in a tube representation. The loops copies with the largest probability are shown in color together with the residues known to interact with the ligand. Hydrogen atoms are not shown for clarity.

These results appear exceptionally good when considering the fact that the experimentally determined NMR structures have an RMSD slightly larger than 1 Å for the ligand bound position. In addition, the ligands maintain the main interactions with the protein seen in the NMR ensemble. For example, as for the NMR structures, the nitro group of the docked NP2 ligand interacts with Y809 through a hydrogen bond with the hydroxyl, and the ammonium group interacts with E750. Similar hydrophobic contacts to the ones observed in the NMR ensemble [4] are also maintained in the docked pose.

The docked NP1 ligand maintains the interaction between the ammonium group and E750 while substituting the nitro group interaction with Y809 for Y802. The higher RMSD (2.49 Å) of this docked pose reflects this shift. This interaction has also been observed for this ligand by Zeng et al. [4] to complement Y809. A survey of the NMR structures for this complex shows that depending on the NMR model either tyrosines can interact with the ligand's nitro group, with Y802 being more frequent than Y809.

CONCLUSIONS

A new protocol was developed for structure-based ligand docking that takes into account the flexibility of the protein binding pocket. The use of multiple copies of the flexible regions in this method is general, but takes special advantage of the ensemble of structures derived from NMR to create multiple copies of the binding pocket; the rest of the protein is represented as a single framework. The energetic coupling of the multiple copies region and the MC/SA sampling is achieved with a mean field theory approximation, and dynamic reassignment of probabilities to each copy. Both side-chain and backbone flexibility is included in the sampling of docking poses, by allowing the side-chains to move while several copies of the backbone with different conformations are kept fixed.

Starting from an ensemble of NMR models of the P/CAF structure the method was shown to be successful in docking four ligands in positions agreeing well with structural data.

ACKNOWLEDGMENTS

This work has been supported by National Institutes of Health awards P01-GM66531 and K05 DA00060. We thank Ming-Ming Zhou and Lei Zeng for sharing the NMR structures of the P/CAF BRDs in complex with NP3 and DMC.

REFERENCES

[1] D. E. Koshland, Jr., W. J. Ray, Jr., and M. J. Erwin, *Fed. Proc.* **17**, 1145 (1958).
[2] S. J. Teague, *Nat. Rev. Drug. Discov.* **2**, 527 (2003).
[3] X. Fradera, X. De La Cruz, C. H. Silva et al., *Bioinformatics* **18**, 939 (2002).
[4] L. Zeng, J. Li, M. Muller et al., *J. Am. Chem. Soc.* **127**, 2376 (2005).
[5] K. Kaehlcke, A. Dorr, C. Hetzer-Egger et al., *Mol. Cell.* **12**, 167 (2003).
[6] C. Dhalluin, J. E. Carlson, L. Zeng et al., *Nature* **399**, 491 (1999).
[7] S. Mujtaba, Y. He, L. Zeng et al., *Mol. Cell.* **9**, 575 (2002).
[8] P. Koehl and M. Delarue, *Curr. Opin. Struct. Biol.* **6**, 222 (1996).
[9] K. Bastard, A. Thureau, R. Lavery et al., *J. Comp. Chem.* **24**, 1910 (2003).
[10] A. Roitberg and R. Elber, *J. Chem. Phys.* **95**, 9277 (1991); C. M. Stultz and M. Karplus, *J. Chem. Phys.* **109**, 8809 (1998).
[11] B. R. Brooks, R. E. Bruccoleri, B. D. Olafson et al., *J. Comp. Chem.* **4**, 187 (1983).
[12] NMR structures for NP3 and DMC provided by M-M. Zhou.
[13] N. Brooijmans and I. D. Kuntz, *Annu. Rev. Biophys. Biomol. Struct.* **32**, 335 (2003).

Sequence and Phylogenetic Analysis of FAD Synthetase

Luisa Schubert, Susana Frago, Marta Martínez-Júlvez, and Milagros Medina

Departamento de Bioquímica y Biología Molecular y Celular, Facultad de Ciencias and Institute of Biocomputation and Physics of Complex Systems, Universidad de Zaragoza, Zaragoza, Spain

Abstract. An evolutionary analysis of the sequences available till now for FAD synthetases has been carried out. Several identical conserved residues have been observed along the sequences of all the FAD synthetases analyzed, which might correlate with role for these residues in the catalytic activity of the enzyme. Phylogenetic analysis shows that FAD synthetase sequences can be organized in two main clusters. One of them mainly contains temperature, pressure or pH resistant organisms, whereas in the other one organisms with pathogenic character can be found.

Keywords: FAD synthetase, Sequence, Phylogeny.
PACS: 87.15.Rn, 36.20.–r, 87.14.Ee

BACKGROUND

Biosynthesis of FMN and FAD from their riboflavin precursors in most prokaryotes is catalyzed by a bifunctional flavin adenine dinucleotide enzyme, FAD synthetase. Conversion of riboflavin into FAD by FAD synthetase involves a phosphorylation reaction followed by subsequent adenylylation of FMN to generate FAD [1]. The FAD synthetase C-terminal region is proposed to be involved in the kinase activity, whereas the N-terminal, which is well conserved and with remote similarity to nucleotidyltransferases, is proposed to be involved in the reaction of adenylylation [2]. Most organism's activity highly depends of the function of several key flavoproteins and flavoenzymes that catalyze a variety of biological redox reactions and that require the FAD and FMN redox cofactors. Therefore, since in many of these organisms FAD synthetase is required for their viability and appears unrelated to mammalian enzymes, this protein might be a potential target for the development of novel antimicrobial drugs. So far, the three-dimensional structure of a FAD synthetase has only been reported for the *Thermotoga maritima* enzyme [3]. In this study, sequence analysis of known bifunctional FAD synthetases is used to provide information about putative key residues in the enzyme mechanism, which is still under discussion, and to suggest possible mutation sites or points for drug target. Additionally, we would like to get information about divergence within the different members of the FAD synthetase family. This will be done by determining proximity of the gene evolution in different species.

CP851, *From Physics to Biology; BIFI 2006 II International Congress,*
edited by J. Clemente-Gallardo, Y. Moreno, J. F. Sáenz Lorenzo, and A. Velázquez-Campoy

STRATEGIES AND ISSUES

An evolutionary analysis of the sequences of the FAD synthetase family is shown.

Sequence Analysis of FAD Synthetase

Over the complete FAD synthetase amino acid sequence of *Corynebacterium ammoniagenes* non-redundant data base PSI-BLAST query at http://www.expasy.ch/tools/blast/ was performed. BLAST query generated 2.752 038 results showing homology with FAD synthetase from *Corynebacterium ammoniagenes*. BLAST results displaying E-values ≤ e^{-38}, a total of 97 sequences, were chosen for alignment. Multiple sequence alignment was performed using the CLUSTAL W algorithm (http://www.ebi.ac.uk/clustalw/) with default parameters. Alignment was visualized and converted into a PHYLIP-compatible file by Bioedit (http://www.mbio.ncsu.edu/BioEdit/page2.html) (Figure 1).

	22 25 26 28 31	123 129	164 168	207 210	223	255	268	292	310
C. ammoniagenes	IGVFDGVHRGHQ	VGANFTFGE	SSTTVRE	FPTANQ	DG-V	SVGT	--SVESFV	VRA	AKDV
T. maritima	IGVFDGVHIGHQ	VGRDFRFGK	SSSLIRN	FPTANI	RG-V	NVGF	-VKYEVYI	MRD	DQDV
A. aeloicus	VGNFDGVHLGHR	VGYDWRYGY	SSTLIRR	FPTANL	EG-V	NYGY	KRVLEVHI	IRE	KKDV
B. subtilis	LGYFDGVHLGHQ	AGFDFTYGK	SSSYIRT	FPTANV	TG-V	NIGY	--SIEVNL	IRS	EKDK
C. perfringens	LGSFDGIHKGHL	VGFNYRFGH	SSTRIRK	FPTANL	IG-V	SVGN	-TTVETHI	IRN	IKDK
C. efficiens	IGVFDGVHRGHQ	VGENFTFGT	CSTLVRE	YPTANL	DG-V	SVGT	--SVEAFV	LRP	ANDV
D. vulgaris	IGNFDGVHQGHR	VGYDFSLGK	SSTRIRD	FPTANI	PG-V	SIGH	-LSVETHI	LRE	RHDV
G. kaustophilus	LGYFDGIHLGHQ	AGFDFTYGR	SSSRVRK	FPTANI	VG-V	NVGY	LPNIEVHL	LRS	ARDK
L. interrogans	LGNFDGIHLGHQ	IGFNHCFGK	SSSYVRR	FPTANV	IG-V	NIGR	SLTLESHI	IRD	KQDE
L. welshimeri	LGFFDGVHLGHQ	AGFDYSYGK	SSTNIRR	FPTANI	LG-V	SIGY	LS-IEVYI	FRP	EKDE
M. tuberculosis	IGVFDGVHRGHA	VGENFTGK	SSTYIRS	FPTANV	DG-V	SVGT	--TVEAFV	IRG	GADT
O. iheyensis	IGFFDGIHQGHQ	AGFDYSFGY	SSTRIRK	YPTANI	QG-I	NIGT	VS-VEVYI	VRP	KEDE
T. tengcongensis	LGNFDGVHIGHQ	VGPNYRFGH	SSSLIRE	FPTANV	RG-V	NVGF	-LSVETHI	IRD	LKDI

FIGURE 1. Regions and residues showing high conservation of identical residues along the FAD synthetase sequences analyzed. *Corynebacterium ammoniagenes* FAD synthetase numbers are used.

Despite in many cases the identity of residues among sequences represents only 30 -40 % (Table 1), sequence alignment displayed an important number of identical residues in (almost) all sequences analyzed, namely: G22, G26, H28, H31, G129, R168, T208, N210, G223, G255, E268, R292 and D310 (*Corynebacterium ammoniagenes* numbering) (Figure 1). However, many other residues are conserved in a large number of the sequences analyzed. Thus, *Norcardioides sp JS614* is the only sequence that presented an arginine (R) instead of aspartate (D) at position denoted as 25, whereas residue 123 (glycine in all other sequences) is replaced by an aspartate (D) in *Norcardioides sp JS614* and by an alanine (A) in *Thermobifida fusca*. Additionally, other positions also displayed conservative substitutions (e.g. G125, S164, T165). Such strong conservation of some residues might be correlated to an indispensable role for the catalytic activity of the enzyme.

TABLE 1. Percentage of identical and homologous residues of different FAD synthetase sequences with regard to *C. ammoniagenes*.

ORGANISM	% IDENTITY	% CONSERVATIVE SUBSTITUTIONS	% GAPS
Corynebacterium ammoniagenes	-	-	-
Thermotoga maritima	32	50	9
Aquifex aeolicus	31	53	5
Bacillus subtillis	30	52	5
Clostridium perfringensis	33	54	5
Corynebacterium efficiens	59	71	2
Desulfovibrio vulgaris	38	55	4
Geobacillus kaustophilus	35	55	6
Leptospira interrogans	30	52	4
Listeria welshimeri	33	52	5
Mycobacterium tuberculosis	45	59	2
Oceanobacillus iheyensis	31	54	5
Thermoanaerobacter tengcongensis	36	57	5

Phylogenetic Analysis of FAD Synthetase

Phylogenetic trees were calculated, evaluated and visualized using the Phylip program package (http://evolution.genetics.washington.edu/phylip.html), version 3.6a3, with the maximum likelihood algorithm -"PROML" as tree-building method [4,5,6]. In order to generate bootstrapped trees, the "SEQBOOT"-function (number of bootstrap trees: 100) was applied upon the original data set [8,9]. With the generated 100 bootstrap-trees the maximum likelihood data-matrix for each of these trees was calculated before a consensus tree was created with the function "CONSENSE". Because all data are mostly provided as mathematical matrices, TREEVIEW (http://taxonomy.zoology.gla.ac.uk/rod/treeview.html) was used to generate a rooted tree to visualize the data. To generate the best possible tree, maximum likelihood algorithm was applied upon the alignment.

The bootstrapped tree or cladogram containing all 97 sequences (not shown) indicate that FAD synthetase sequences of prokaryote families (such as *Mycobacteria* or *Corynebacteria*) are clustered, suggesting fidelity of the method. The 97-sequences-tree was reduced in number of species to maximize clarity. 17 redundant sequences (*Synechococcus, Listeria monocytogenes, Listeria welshimeri, Trypheryma whipplei, Frankia, Xanthomonas campestris, Zymomonas mobilis, Leptospira interrogans* seroformes) were deleted out of the alignment and a second phylogenetic procedure was performed generating a 80-sequences-containing tree (Figure 2; bootstrap-values not shown). Sequence-deletion did not cause any information-loss because one member of each cluster-group was always kept in the new tree representing thus deleted ones. The new tree was generated in a time-consuming but accurate way (different PHYLIP parameters).

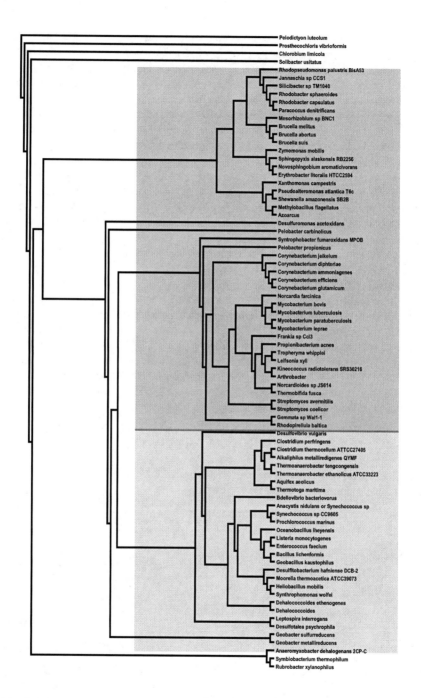

FIGURE 2. Phylogenetic relationships in FAD synthetases. Clusters are indicated by different grey background and separation between them is shown by a darker grey line.

183

In the generated tree, two main clusters can be distinguished. Interestingly the cluster (lower one, light grey) containing the *Thermotoga maritima* sequence, also includes several other temperature, pressure or pH resistant organisms. Thus, this cluster contains bacillales such as *Oceanobacillus, Bacillus, Geobacillus and Listeria,* the clostridia, *Moorella, Thermoanaerobacter, Syntrophomonas, Heliobacillus, Alkaliphilus* and *Clostridium,* the deltaproteobacteria *Desulfovibrio, Geobacter, Desulfotalea, Bdellvibrio* and *Syntrophomonas,* the aquificae *Aquifex,* the chloroflexi *Dehalococcus* and, also some cyanobacteria such as *Prochlorococcus* and *Synechococcus.*

The other cluster (upper one, dark grey) contains mainly organisms that present some pathogenic members such as the actinobacteriae *Corynebacteria, Mycobacterium, Propionibacterium, Thermobifida, Arthrobacter, Kineococcus, Frankia* or *Nocardiodes,* the alphaproteiobacteria *Brucella, Rhodopseudomonas, Zymomonas* or *Rhodobacter* as well as the planctomycetes *Gemmata* and *Rhodophirellula.* This clustering suggests than from the phylogenetic point of view FAD synthetase diverged to form two big groups, which relate quite well to the evolution of the bacteria hosting them.

Therefore, sequence comparison and evolution analysis here presented suggest that the structure reported for *Thermotoga maritima* FAD synthetase might be an adequate model for enzymes of its same cluster. However, the sequence and phylogenetic divergences shown between *Thermotoga maritima* FAD synthetase and the enzymes from pathogenic organisms, such as *Mycobacterium* or *Corinebacterium,* will probably be reflected in structural divergences that might affect the catalytic mechanism and might make possible the design of specific antimicrobial drugs for the enzymes of these organisms.

ACKNOWLEDGMENTS

This work has been supported by Grant BIO2004-00279 to M.M.. L.S, thanks the Socrates-Erasmus program for a Fellowship.

REFERENCES

1. J. Manstein and E.F. Pai, *J. Biol. Chem.* **261**, 25-30 (1986).
2. A. Krupa, K. Sandhya, N. Srinivasan and S. Jonnalagadda, *TRENDS in Biochemical Sciences* **28**, 9-12 (2003).
3. W. Wang, R. Kim, J. Jancarik, H. Yokota, S.H. Kim, *Proteins* **52**, 633-635 (2003).
4. J. Felsenstein, *Am. J. Hum. Gen.* **25**, 471-492 (1973).
5. J. Felsenstein, *Evolution* **39**, 783-791 (1985).
6. W. M. Fitch, Systematic Zoology **20**, 406-416 (1971).
7. M. Kimura, *J. Mol. Evol.* **16**, 111-120 (1980).
8. H. Kishino and M. Hasegawa, *J. Mol. Evol.* **29**, 170-179 (1989).
9. C.F.J. Wu, *Annals of statistics* **14**, 1261-1350 (1986).

Tissue-specific *Ctr1* Gene Expression and *in silico* Analysis of Its Putative Protein Product

Sergey A. Samsonov*, Eija Nordlund[†], Natalia A. Platonova[¶], Alexey N. Skvortsov*, Nadezhda V. Tsymbalenko[¶], Ludmila V. Puchkova*,[¶]

*Biophysical Department, Saint-Petersburg State Polytechnical University, 195251 Polytechnicheskaya street 29, Saint-Petersburg, Russia.
[¶]Department of Molecular Genetics, Research Institute for Experimental Medicine (RAMS), 197376, Pavlov street 12, Saint-Petersburg, Russia.
[†] Turku Centre for Computer Science, Lemminkäisenkatu 14 A, FIN-20520, Turku, Finland

Abstract. Investigations of the links between *Ctr1* gene activity and copper status in rat organs (liver, cerebellum, choroid plexus and mammary gland) with distinct types of copper metabolism as well as theoretical analysis of CTR1 domains structure were carried out in the research. The results suggest that *(i)* activity of mammalian *Ctr1* gene is tissue-specific regulated at least by two different mechanisms: the gene activity is repressed by high intracellular Cu content and is activated/inactivated dependently on the cuproenzymes synthesis level required by physiological conditions. *(ii)* Multimerized conservative transmembrane domains 2 and 3 form the channel with copper binding amino acid side chains groups oriented inside this channel. These groups can transfer copper to the cytosolic domain, where Cu binds to CTR1 cytosolic HCH-motifs and can be further transferred to CXXC-motif of any known Cu(I)-chaperon.

Keywords: *Ctr1* gene expression, copper metabolism in ontogenesis, molecular modeling.
PACS: 01.30.Cc

INTRODUCTION

Copper (Cu) is a cofactor for many vital enzymes, at the same time copper is highly toxic when protein-unbound or present in excess of cellular needs [1]. High affinity Cu transporter CTR1 (encoded by *Ctr1, Entrez SLC31A1*) is the main candidate for the role of Cu(I)-importer in mammals [2]. *Ctr1[-/-]* mice died early *in utero*. In *Ctr1[+/-]*, Cu content in brain and spleen decreased, while Cu concentration in liver remained at the same level [3]. Besides, the gene expression did not depend on dietary Cu uptake [4]. Neither donor nor acceptor of Cu for CTR1, nor CTR1 intimate function is known. The further study of *Ctr1* gene function is required for a deeper understanding of molecular mechanisms underlying the functioning of copper metabolic system [5], even slight disruptions in which lead to the development of severe neurodegenerative diseases [6]. It was also shown that CTR1 plays crucial role in the specific uptake of cisplatin, antitumour platinum drug [7]. The aim of this research was the search for the relations between *Ctr1* activity and copper status in rat organs with distinct types of copper metabolism and the theoretical analysis of CTR1 domain structure. In the experimental part of the research the rat liver, cerebellum and choroid plexus of the embryonic and adult copper metabolism types, as well as mammary gland during lactation were used. Copper status was defined as copper concentration and the activity of ceruloplasmin (Cp) gene that encodes two protein isoforms of cop-

CP851, *From Physics to Biology; BIFI 2006 II International Congress,*
edited by J. Clemente-Gallardo, Y. Moreno, J. F. Sáenz Lorenzo, and A. Velázquez-Campoy
© 2006 American Institute of Physics 0-7354-0350-3/06/$23.00

per inducible multifunctional blue ferroxidase (secretory Cp and glycosil phosphatidyl inositol anchored isoform, GPI-Cp) formed by the mechanism of alternative splicing [8, 9].

METHODS

Total RNA was extracted with TRIZOL Reagent ("Roche", USA). *RT-PCR was* performed with the use of the reagents by "Promega" (USA). Primers for *Ctr1* (NM_13360: T^{287}-G^{308}, A^{600}-G^{620}), *Cp* (NM_012532: A^{1966}-T^{1990} and T^{3202}-A^{3226}), and *GPI-Cp* (NM_012532: A^{1966}-T^{1990} and AF202115: C^{3228}-A^{3252}); *β-actin* (NM_031144: G^{651}-G^{670} and A^{958}-G^{977}) were used. Fractional amplification method was used [10]. *Copper concentration* was measured by atomic absorption spectrometry with Zeeman correction («Perkin-Elmer», Model 4100ZL, USA). *Cp protein concentration* was defined by rocket immunoelectrophoresis. For *in silico analysis* CTR1 amino acids (aa) sequences were retrieved from NCBI. ClustalW, GOR4, PsiPred, package PHYLIP 3.6 [11], RasMol 2.7.2.1 [12] software was used. Geometry optimization in HyperChem 7.01 package [13] was performed by the molecular mechanics method MM+ with the atomic charges calculated by the quantum-chemical semi-empirical method ZINDO/1. The chosen methods handle *d*-shell elements, solvent was not accounted for. The upstream regions of *Ctr1* genes of different species were analyzed by the programs Multalin, MultTF, and the package GEMS Launcher.

RESULTS

The experimental data on the expression of *Ctr1* and copper status are summarized in Figure 1. In liver CTR1-mRNA level increased until reaching certain Cu concentrations in nuclei, then drastically decreased up to the 13th day of life while Cu content in lysosomes progressively rose (Fig. 1a,b). In the adult copper metabolism type CTR1-mRNA content increased accordingly to the increase of Cp-mRNA level (Fig. 1b). At the same time soluble Cp and Cu content in blood changed from 12.2±4.3 to 35.5 ± 3.1 (mg/100 ml) and from 230 ± 21 to 900 ± 31 (ng/mg of protein), respectively. Cu concentration rose in cerebellum during the first month of life and increased twice (Fig. 1c). *Ctr1* activity changed similarly (Fig. 1d). In cerebellum, both Cp and GPI-Cp mRNAs were found. However, at the embryonic copper metabolism type soluble Cp-mRNA isoform was predominant, while at the adult copper metabolism type GPI-Cp isoform prevailed (Fig. 1d). In choroid plexus copper status changed in different manner during development (Fig. 1c), similarly to liver. Here, Cu content increased after the birth and then dropped. In parallel, CTR1-mRNA content (Fig. 1d) was low under the high Cu content and increased twofold after intracellular Cu concentration decrease. In adults, both Cp-mRNA forms levels increased. In mammary gland CTR1-mRNA level increased after delivery and decreased during lactation (Fig. 1f). After delivery Cp-mRNA content rose almost tenfold and then gradually fell. In the skimmed milk Cp and Cu concentrations accordingly decreased during lactation (Fig. 1e). GPI-Cp mRNA level slowly decreased since the last days of pregnancy and was not associated with Cu content changes in milk.

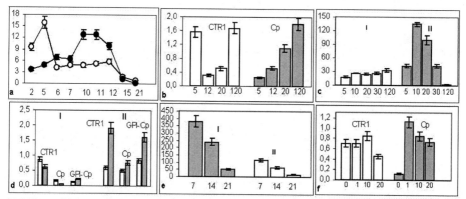

FIGURE 1. Tissue-specific links between rat *Ctr1* gene expression and copper status in ontogenesis.
  (a) Cu distribution in nuclei (white) and lysosomes (black) of liver cells during switching of Cu metabolism type. Abscissa: age, days; ordinate: Cu concentration, ng/mg of total protein. (b) Steady state CTR1-mRNA (white) and Cp-mRNA (grey) levels in liver in development. Abscissa: age, days; ordinate: mRNA content, aperture units, a. u. (c) Cu concentration in cerebellum (I) and choroid plexus (II) during change of Cu metabolism type. Abscissa: age, days; ordinate: Cu concentration, ng/mg of total protein. (d) Steady state of CTR1, Cp, GPI-Cp mRNA levels in cerebellum (I) and choroid plexus (II) in 10- (white), 120-day olds (grey). mRNA content, a. u. (e) Skimmed milk Cu (grey) and Cp (white) concentration during lactation. Abscissa: days after delivery; ordinate: Cu concentration, μg/L; Cp concentration, mg/L. (f) Steady state of CTR1 (white) and Cp (grey) mRNA levels in mammary gland. Abscissa: days after delivery; ordinate: mRNA content, a. u. The average values were calculated in 3 independent experiments for Cp and Cu concentrations and mRNA content measurements.

The obtained data allow us to deduce two types of the relations between *Ctr1* expression and copper status: *(i)* CTR1-mRNA level decreases under Cu accumulation in nuclei; *(ii)* *Ctr1* activity is interconnected with the level of cuproenzymes. It is possible to conclude that transcriptional activity of mammalian *Ctr1* gene is regulated tissue-specifically by at least two different mechanisms. These data on organ-specific *Ctr1* expression agrees well with other research data showing that the CTR1 protein was found in liver, lungs, kidneys, heart, brain, spermatocytes, muscles, and erythrocytes and its level changed in ontogenesis or under distinct copper status [14]. To reveal the transcription mechanisms underlying tissue-specificity we predicted transcription factor binding sites (TFBSs). For this sake *Ctr1* genes 3000 bp upstream region sequences were extracted from *Entrez* nucleotide database. TFBSs prediction in the *cis*-regulatory region of human *CTR1* gene was accomplished by the methods of comparative genomics. We searched for the TFBSs that are conservative in different species (human, rat and mouse) and present in clusters. General TFBSs required for eukaryotic transcription were found as well as the ones considered to regulate differentiation of central neural system, hepatocytes, lymphoid cells and others. Besides, we found TFBS for SSRP1 sequestered by cisplatin treatment. The presence of these TFBSs agrees with the functions of CTR1 previously described by use of different experimental and theoretical approaches. Noticeably, no metal responsive elements were found in *cis*-regulatory sequence of the gene.

CTR1 functioning is still unclear, however, the knowledge about this protein structural organization and molecular mechanisms of copper binding could be useful in understanding of CTR1 role in copper metabolism. Theoretical analysis is one of the tools to reach this aim. That is why we utilized different *in silico* approaches.

The membrane topology model of CTR1 is similar for all organisms and includes N-extracellular, three transmembrane (TMD) and C-cytosolic domains [2]. Phylogenetic analysis of CTR1 sequence and its topological domains revealed that TMD2, TMD3 are the most conservative (Tabl. 1). N-terminal sequences are the most variable between yeast and high eukaryotes but all of them are capable of copper binding [15]. Perhaps, this difference is related to the distinct copper sources for the uni- and multicellular organisms.

TABLE 1. Phylogenetic distance between CTR1 sequences (Kimura model).

	Complete CTR1	N-terminus	TMD1	TMD2	TMD3	C-terminus
Mammals	0.107	0.343	0	0	0	0
Vertebrates	0.271	0.746	0.294	0.196	0.031	0
Other species	2.641	2.798	2.150	1.418	1.909	2.390

It was shown that mammalian CTR1 predominantly exists in membrane as homotrimer [16]. Based on these data, we tried to propose the model of the CTR1 channel. Predicted secondary structure of TMD1-3 strongly tends to α-helical conformation. Probable orientation of aa residues in TMDs of CTR1 is shown in Fig. 2a. TMD1 has no potential Cu binding sites and is entirely hydrophobic, while TMD2 and TMD3, that theoretically able to form a hydrophilic surface, contain 21 aa side chain groups able to coordinate Cu and orient inside the predicted channel. Considering hydrophilicity of the TMDs, the channel formed by all TMDs [2, 17] (Fig. 2b) is less likely to exist than by TMD2 and TMD3 of homotrimer (Fig. 2c). Since the radius of α-helix is 2.3 Å, the radius of the channel formed by 9 TMDs is no less than 6.6 Å, while the radius of the channel formed by 6 TMDs is about 4.4 Å. The recently published data on 2D CTR1 crystallization shows that experimentally defined channel transmembrane pore has formal C_3 symmetry and its radius equals 4.5 Å [18]. Thus, our theoretically proposed model fits the experimental data very well. Moreover, the most conservative CTR1 domains TMD2 and TMD3 (Table 1) in yeast also contain 21 aa residues, capable of Cu coordination, oriented inside the channel.

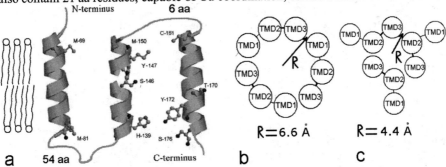

FIGURE 2. Model of CTR1 cuprophilic channel. a – polar and potentially Cu coordinating residues in TMDs hCTR1; b, c - probable TMDs positions in CTR1 homotrimer (view in the plane of membrane).

All the vertebral CTR1 C-terminal cytosolic domains consist of 14 conservative aa, containing the terminal HCH-motif, which coordinates Cu [6]. We estimated *in silico* its Cu binding ability. In the calculated models histidines with deprotonated imidazoles were used, which is relevant to cytosolic pH, and N-termini of HCH-motifs were blocked by N-CH₃ groups. The geometry optimization of HCH-motif in the presence of Cu(I) revealed the key role of His imidazole rings in Cu coordination, while C189 did not participate in the complex formation (Fig. 3).

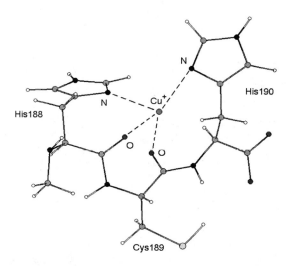

FIGURE 3. Cu(I) coordination by *N*-CH₃-His-Cys-His. Geometry optimization in MM+ with the atomic charges calculated in ZINDO/1.

The results presented in Table 2 show that the substitution of cysteine with serine, methionine, alanine, and lysine changed neither the complex geometry nor the binding energy defined by three methods (MM+, ZINDO/1 and *ab initio* STO-3G*). This agrees with the biochemical data that C189S did not affect CTR1 functional activity [19].

TABLE 2. Binding energy of Cu(I) coordinated by molecule *N*-CH₃-His-X-His.

Method of calculation	Binding energy of coordination complex with Cu(I), kcal/mol		
Molecule	*ab initio* (STO-3G* basis)	ZINDO/1	MM+
N-CH₃-His-Cys-His	277	532	112
N-CH₃-His-Ser-His	270	524	105
N-CH₃-His-Ala-His	278	551	107
N-CH₃-His-Met-His	276	562	116
N-CH₃-His-Lys-His	276	544	100

In the proposed dimer and trimer models of the channel the cytosolic domains are long enough to allow spatial association of HCH-motifs. We modeled HCH-motifs interactions depending on the distance between their N-termini. In the absence of Cu(I) HCH-motifs are favourably spatially remote since the energies of the dimer and trimer systems decreased by about 20 and 50 kcal/mol, respectively, as defined by MM+ method, when the distance between N-termini of the motifs was changed from 3 Å and 4 Å to virtual infinity (thousands of angstroms). This means that in the absence of copper the channel is open at the cytosolic side. The introducing of Cu(I) to the system lead to the complex formation, where N-imidazolic atoms played crucial role. These complexes were more stable than monomeric ones, while trimeric complexes were more stable than dimeric ones (Fig. 4).

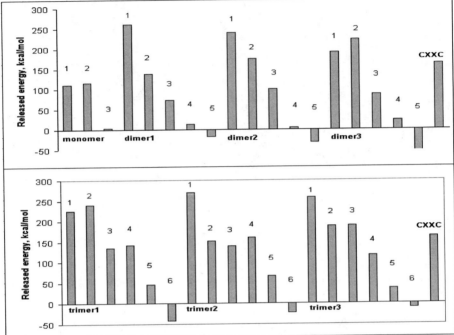

FIGURE 4. The energies released after each consequent copper ion binding (the number of copper ions is shown above each column) compared to the energy released after Cu(I) binding by the CXXC-motif. Dimer 1-3, trimer 1-3 are related to three different symmetric mutual dispositions of HCH-motifs at the initiation of geometry optimization.

The spatial proximity of monomers in the presence of Cu(I) is energetically favoured, the channel is closed. The results obtained by the calculations in ZINDO/1 method qualitatively agree with MM+ results. According to MM+ calculations, to be energetically favoured bound, the maximum number of Cu(I) was 3, 5 and 6 for monomer, dimer, and trimer, respectively (Fig. 4). The complexes of oligomeric HCH-motifs were compared by the calculated energy with cuprophilic CXXC-motif of Cu(I)-chaperons, which are able to coordinate a single Cu(I) by cysteine SH-groups [1KVJ]. The calculations in MM+ showed that HCH-motif with even one Cu(I) bound was able to transfer it to CXXC-motif. As for dimeric or trimeric motifs, they should have at least 2 or 3 Cu(I) already bound to be able to transfer Cu to chaperone (Fig. 4). Perharps, that gives a clue for understanding the fact that CTR1 monomers and homodimers in vitro detected could keep biological function [16].

The applied experimental and theoretical approaches allow us to draw the novel conclusions about CTR1 function. Mammalian CTR1 expression is tissue-specific. TMD1 of CTR1 homotrimer does not participate in the cuprophylic channel formation. Cu(I), after being accepted from extracellular copper transporters (soluble Cp or (His)$_2$Cu complex), is transported through the channel and accepted by C-terminal HCH-motif, which in turn serves as universal copper donor for any known cytosolic Cu(I)-chaperon.

ACKNOWLEDGMENTS

The research was supported by RFBR (06-04-49597)

REFERENCES

1. K. Karlin, *Science* **261**, 701-707 (1993).
2. P.A. Sharp, *Intern. J. Biochem.Cell Biol.* **35**, 288–291 (2003).
3. J. Lee et al., *Proc. Natl. Acad. Sci. U.S.A.* **98**, 6842–6847 (2001).
4. J. Lee et al., *Gene* **254**, 78-96 (2000).
5. S. Puig and D.J. Thiele, *Current Opinion in Chemical Biology* **6**, 171–180 (2002).
6. H. Shim and Z.L. Harris, *J. Nutr.* **133**, 1527–1531 (2003).
7. R. Safaei and S.B. Howell, *Crit Rev Oncol Hematol.* **53**, 13-23 (2005).
8. Patel et al., *J. Biol. Chem.* **275**, 4305–4310 (2000).
9. P. Bielli amd L. Calabrese, *Cell Mol Life Sci.* **59**, 1413-1427 (2002).
10. M. Marone et al., *Biol. Proced. Online* **3**, 19–25 (2001).
11. J. Felsenstein, PHYLogeny Inference Package version 3.57c, *PHYLIP*, 1995.
12. S. Roger, Rasmol Version 2.7.1., *RasWin Molecular Graphics*, 2001.
13. HyperChem 7.0, *Hypercube Inc.*, 1999.
14. Y.M. Kuo et al., *J. Nutr.* **136**, 21-26 (2006).
15. Y. Guo et al., *J. Biol. Chem* **279**, 17428-17423 (2004).
16. J. Lee et al., *J. Biol. Chem* **277**, 4380-4387 (2002).
17. J.F. Eisses and J.H. Kaplan, *J Biol Chem.* **280**, 37159-37168 (2005).
18. S.G. Aller and V.M. Unger, *Proc Natl Acad Sci U S A.* **103**, 3627-3632 (2006).
19. J.F. Eisses and J.H. Kaplan, *J. Biol. Chem* **277**, 29162–29171 (2002).

A Particle Based Implicit Solvent Model for Biomolecular Simulations

Nathalie Basdevant, Tap Ha-Duong and Daniel Borgis

Laboratoire d'Analyse et Modélisation pour la Biologie et l'Environnement - UMR 8587, Université d'Evry Val d'Essonne, Bd François Mitterrand, 91025 Evry Cedex, FRANCE

Abstract. We present a recently developed alternative solvent model for biomolecules simulations that combine advantages of both explicit models (molecular aspect of water for structural informations) and implicit approaches (efficient and rapid calculation of solvation free energies). This model, named *Polarizable Pseudo-Particles (PPP)*, allows stable molecular dynamics simulations in the nanosecond range and yields free energies in good correlation with Poisson-Boltzmann calculations.

Keywords: Solvation, Molecular Dynamics, Electrostatic Free Energy, Proteins, Nucleic Acids, Hydration Sites.
PACS: 87.15.Aa, 87.14.Ee, 87.14Gg.

INTRODUCTION

The solvent plays a very important role in the stabilization of the structures of biomolecules and in their interactions within the cell. In the field of molecular modeling of biological macromolecules, the representation of the solvent is therefore a major issue. We recently developed a semi-implicit solvent model that lies between explicit models, which represents all-atom water molecules, and implicit models, which represents the solvent as a dielectric continuous medium. In this approach, the solvent is considered as a fluid of coarse-grained particles which have macroscopic electrostatic properties, and it results in a fluid of *Polarizable Pseudo-Particles (PPP)*. The PPP model can therefore combine advantages of both methods, keeping a molecular aspect of water and providing fast and accurate calculations of electrostatic properties, in particular solvation free energies.

We present here applications of the PPP model to molecular dynamics and hydration of nucleic acids and proteins of several proteins and nucleic acids, including the BPTI protein, the B1 domain of protein G, the anticodon loop of a tRNA molecule, and both B-DNA and A-DNA molecules.

METHODOLOGY: THE PPP SOLVENT MODEL

The methodology of the PPP solvent model has been described in details previously [1, 2], and we remind here its principal features.

Each water molecule is considered as a coarse-grained Lennard-Jones particle of molecular size that embed a polarizable electric dipole. From the classical macroscopic laws of dielectric materials [3], the dipole $\mathbf{p}_i$ on each particle i can be expressed as a

CP851, *From Physics to Biology; BIFI 2006 II International Congress*,
edited by J. Clemente-Gallardo, Y. Moreno, J. F. Sáenz Lorenzo, and A. Velázquez-Campoy
© 2006 American Institute of Physics 0-7354-0350-3/06/$23.00

function of the vacuum electric field created by the solute partial charges $\mathbf{E}_i^0$, following the relation:

$$\mathbf{p}_i = \frac{\mu}{E_i^0} \mathscr{L} \left(\frac{3\alpha E_i^0}{\mu} \right) \mathbf{E}_i^0, \tag{1}$$

where μ is a phenomenological saturation via a Langevin function $\mathscr{L}(x) = \coth(x) - 1/x$. The introduced polarizability constant, α, is related to the solvent density ρ and to the macroscopic dielectric constant ε by the relation:

$$\alpha = \frac{\varepsilon_0(\varepsilon - 1)}{\rho\varepsilon}, \tag{2}$$

and it includes both electronic and orientational polarization [1]. The electrostatic part of the solvation free energy depends only on the vacuum electric field created by the atomic charges of the solute, and can be simply expressed by:

$$\Delta F_{sol}^{ele} = -\frac{\mu^2}{3\alpha} \sum_i \ln \left[\frac{\sinh(3\alpha E_i^0/\mu)}{3\alpha E_i^0/\mu} \right]. \tag{3}$$

The PPP solvent model has been implemented in the molecular dynamics algorithm ORAC [4] by replacing the solute-solvent Coulomb interactions by the electrostatic solvation free energy in the potential energy function. Parameters of the solvent pseudo-particles were optimized for the simulations of biomolecules [2], and were therefore set to $\mu = 1.5$ Debye for the saturation dipole, $\alpha = 2.33$ Å^3 for the polarizability[1], and $\sigma_{LJ} = 2.88$ Å and $\varepsilon_{LJ} = -3.197$ kJ/mol for their Lennard-Jones parameters.

RESULTS: PROTEINS AND NUCLEIC ACIDS TRAJECTORIES

A set of two proteins[2] and four nucleic acids molecules[3] were simulated for 2 to 4 nanoseconds in a box of PPP solvent molecules, in the canonical NVT ensemble, using Amber94 force-field for biomolecules [5]. Explicit sodium ions were placed near the nucleic acids phosphate groups in order to neutralize the systems.

Molecular dynamics simulations with the PPP solvent model yield stable trajectories in the nanosecond range, with heavy atoms RMSD from starting structures ranging from 1.5 to 2.5 Å [2, 6]. The three-dimensional structures of the simulated proteins, DNA and RNA are in good agreement with simulations in explicit solvent and experimental observations.

---

[1] which corresponds by Eq. (2) to the dielectric constant of water $\varepsilon = 80$ and a density of $\rho = 0.00337$ Å^3.

[2] PDB codes: 5pti and 1pgb.

[3] From the NDB, a B-DNA (bdl001), the same DNA molecule both in A form (adj109) and B form (bd0015), and the anticodon loop of a tRNA molecule (trna05).

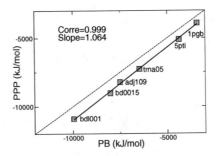

FIGURE 1. Electrostatic solvation free energies for the studied proteins and nucleic acids. Comparison between Poisson-Boltzmann calculations (PB) and values obtained with the PPP solvent model [6].

Electrostatic Solvation Free Energies

Electrostatic solvation free energies were estimated "on-the-fly" with the PPP solvent model, according to Eq. (3), and compared to the reference values calculated with an implicit solvent model by resolution of the Poisson-Boltzmann equation.

For each biomolecule, variations of the electrostatic solvation free energy due to solute conformational changes correlate well with the Poisson-Boltzmann variations. Moreover, for all tested biomolecules, as shown on Fig. 1, correlations between averages of electrostatic solvation free energies estimated with the PPP model and Poisson-Boltzmann calculations are very satisfactory (correlation coefficient and slope are very close to one).

Hydration Sites of Nucleic Acids

In accordance with experimental observations, as shown on Fig. 2, molecular dynamics simulations in the PPP solvent show that A-DNA are preferentially hydrated in the major groove and B-DNA in the minor groove.

Despite the absence of explicit hydrogen bonds of the solvent pseudo-particles, most of the experimentally identified preferential hydration sites at the nucleic acids surfaces are reproduced during the simulations.

CPU Times

Table 1 shows the comparison of CPU Times between the PPP solvent and an explicit solvent model, TIP3. For a solute with about 1000 atoms, the CPU gain of the PPP solvent with respect to TIP3 is about 5, proving that the PPP solvent model is very efficient.

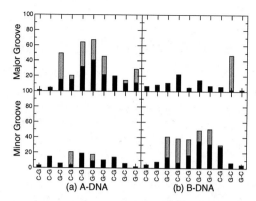

FIGURE 2. Maximum residence time (% of simulation length) of solvent (black bars) and ions (grey) near polar groups in the major (upper panel) and minor grooves (lower) of a double-helix of DNA in A form (a) and B form (b) [6].

TABLE 1. Comparison of CPU times for trajectories of solutes in a box of solvent molecules PPP and TIP3 [2].

Solute	Solute Atoms Number	Solvent Molecules Number	Ratio TIP3/PPP
Protein (1pgb)	855	5202	4,5
B-DNA (bdl001)	782	5169	5,0
tRNA (trna05)	562	5248	5,4

CONCLUSION

Remembering that the PPP model efficiently provides solvation free energies as well as preferential hydration sites on solutes surfaces, it allows future studies of solvation influence on recognition processes between biomolecules, either at an all-atom or coarse-grained level (with larger and softer solvent polarizable pseudo-particles).

ACKNOWLEDGMENTS

N.B. is supported by a grant from Genopole®, which is gratefully acknowledged.

REFERENCES

1. T. Ha-Duong, S. Phan, M. Marchi, S. Phan, and D. Borgis, *J. Chem. Phys.* **117**, 541–556 (2002).
2. N. Basdevant, D. Borgis, and T. Ha-Duong, *J. Comp. Chem.* **25**, 1015–1019 (2004).
3. D. F. Calef, and P. G. Wolynes, *J. Chem. Phys.* **87**, 3387–3399 (1983).
4. P. Procacci, T. A. Darden, E. Paci, and M. Marchi, *J. Comp. Chem.* **18**, 1848–1862 (1997).
5. W. D. Cornell, P. Cieplak, C. I. Bayly, I. R. Gould, K. M. Merz, D. M. Ferguson, D. C. Spellmeyer, T. Fox, J. W. Caldwell, and P. A. Kollman, *J. Am. Chem. Soc.* **117**, 5179–5197 (1995).
6. N. Basdevant, T. Ha-Duong, and D. Borgis, *J. Chem. Theor. Comp.* (submitted).

Mechanism of Metal Ion Activation of the Diphtheria Toxin Repressor DtxR

J. Alejandro D'Aquino[1] and Dagmar Ringe[1]*

[1]*Departments of Chemistry and Biochemistry and Rosenstiel Basic Medical Sciences Research Center, Brandeis University, Waltham, Massachusetts 02454.*

Abstract

The diphtheria toxin repressor, DtxR, is a metal ion-activated transcriptional regulator that has been linked to the virulence of *Corynebacterium diphtheriae*. Structure determination has shown that there are two metal ion binding sites per repressor monomer, and site-directed mutagenesis has demonstrated that binding site 2 (primary) is essential for recognition of the target DNA repressor, leaving the role of binding site 1 (ancillary) unclear (1 – 3). Calorimetric techniques have demonstrated that while binding site 1 (ancillary) has high affinity for metal ion with a binding constant of 2×10^{-7}, binding site 2 (primary) is a low affinity binding site with a binding constant of 6.3×10^{-4}. These two binding sites act independently and their contribution can be easily dissected by traditional mutational analysis. Our results clearly demonstrate that binding site 1 (ancillary) is the first one to be occupied during metal ion activation, playing a critical role in stabilization of the repressor. In addition, structural data obtained for the mutants Ni-DtxR(H79A,C102D), reported here and the previously reported DtxR(H79A) (4) has allowed us to propose a mechanism of metal ion activation for DtxR.

Introduction

Metal ion activation is one of the central issues in understanding repression of virulence genes by DtxR. While DtxR is activated by ferrous ion *in vivo*, DtxR could be activated by other divalent transition metal ions *in vitro* with the following order of activation: $Fe^{2+} \sim Ni^{2+} > Co^{2+} >> Mn^{2+}$ (5,6) Early binding experiments showed that DtxR has a high-affinity metal ion binding site with an affinity for Ni^{2+} between 2.11×10^{-6} and 9×10^{-7}(7,8). Mutagenesis experiments have shown that alanine substitution of any of the ligands of binding site 2 (primary) results in an inactive mutant that no longer binds to its operator DNA(1,2). The interpretation of similar experiments on binding site 1 (ancillary) has not been as clear. Early mutational studies on this site show small decreases in activity and suggest that it is expendable(1). More recent *in vivo* studies determined that a minimum of two alanine substitutions in this site are required to abolish activity of the repressor(2).

In order to elucidate the individual roles of each of the binding sites and the mechanism by which activation occurs, we used crystallographic and calorimetric techniques to study a self-consistent set of repressors, including WT, single mutants in each of the binding sites and a double mutant. The effects of these mutations on structure

CP851, *From Physics to Biology; BIFI 2006 II International Congress,*
edited by J. Clemente-Gallardo, Y. Moreno, J. F. Sáenz Lorenzo, and A. Velázquez-Campoy
© 2006 American Institute of Physics 0-7354-0350-3/06/$23.00

and binding affinity have been determined as well as potential cooperative effects between the binding sites. Comparison of the crystal structure of Ni-DtxR(H79A,C102D), with those of wild type DtxR, site 1 mutant DtxR(H79A) and site 2 mutant Ni-DtxR(C102D), correlate conformational changes with differences in metal ion coordination at both binding sites. We have used differential scanning calorimetry (DSC) to study the thermal stability and oligomerization state of DtxR in the presence and absence of metal ion. Our results have allowed us to isolate important events during metal ion activation of DtxR and to propose a molecular mechanism that fits all the available data.

Results and Discussion

While the structure of DtxR(H79A) is superimposable with the metal-free form of wild type DtxR, both lacking metal ion in both binding sites, the structure of Ni-DtxR(H79A,C102D) has metal ion bound to site 2 (primary) only, and has no other structural changes than those resulting from the H79A mutation. We have concluded that metal ion binding site 2 is solely responsible for the conformational changes associated with activation. The N-terminal region of DtxR(H79A) does not show any of the conformational changes associated with activation by metal ion.

Binding affinities

We have obtained the occupancy and relative binding affinities of each of the binding sites using ITC. The titration of DtxR and all the mutants used in this study with Ni^{2+} is best described by a model consisting of two binding sites, one of high affinity and one of low affinity. Wild type DtxR has a high affinity (1.7×10^{-7}) and a low affinity binding site (6.4×10^{-4}). Replacement of His by Ala in DtxR(H79A), lowers the binding affinity of the high affinity binding site by one order of magnitude but does not affect the affinity of the low affinity binding site beyond experimental error. Similarly, replacement of Cys 102 by Asp in DtxR(C102D) shows a decrease of one order of magnitude in the binding affinity of the low affinity binding site. The binding affinitiy of DtxR(H79A,C102D) decreases approximately by one order of magnitude for each of the binding sites. From the previous discussion, it could be inferred that binding site 1 is the high affinity binding site and binding site 2 the low affinity binding site. The fact that binding site 1 (ancillary) has the higher affinity for metal ion indicates that the active repressor has a fully occupied binding site 1 (ancillary). Despite the decreased binding affinity at binding site 1 DtxR(H79A) remains active *in vivo*. In fact, *in vivo*, a minimum of two alanine substitutions in binding site 1 (ancillary) are required to abolish activity of the repressor(2), and therefore, the role of binding site 1 in the function of DtxR cannot be underestimated. In addition, since our results show the effect of these mutations on the metal-ion binding affinity to be simply additive, the possibility of cooperativity between the two binding sites would seem to be ruled out.

Thermal stability of DtxR

In order to study the effect of the metal ion on the oligomerization state of DtxR, we measured its thermally induced folding-unfolding transitions. In the range of concentrations studied, the transition temperature of metal-free DtxR(C102D) is independent of the protein concentration, indicating that the protein behaves as a monomer. Incubation of the protein sample with various concentrations of $NiCl_2$ leads to aggregation prior to the transition temperature.

A Working Model

The combined use of calorimetric and crystallographic techniques presented in this work has increased our understanding of the role of each metal ion binding site at the molecular level and has provided a clearer picture of the order in which different events take place during metal ion activation of DtxR. The activation mechanism can be summarized as follows: at low concentrations of metal ion, DtxR exists essentially as a monomer. As the concentration of metal ion increases, binding site 1 (ancillary) becomes occupied and stabilizes the protein. There are no significant conformational changes in the N-terminal region of the repressor associated with metal ion binding to site 1 (ancillary). Once binding site 1 (ancillary) is occupied, metal ion is able to bind to binding site 2 (primary), leading to small conformational changes in the N-terminal region; the first 6 residues of N-terminal helix unravel allowing the water-mediated interaction between the carbonyl oxygen of Leu4 and binding site 2 (primary) metal ion to occur. Once the repressor is activated, the repressor is able to interact with its operator DNA.

References

1. Ding, X., Zeng, H., Schiering, N., Ringe, D. & Murphy, J. R. (1996) *Nature Structural Biology* **3**, 382-387.
2. Goranson-Siekierke, J., Pohl, E., Hol, W. G. J. & Holmes, R. K. (1999) *Infection and Immunity* **67**, 1806-1811.
3. Love, J. F., vanderSpek, J. C. & Murphy, J. R. (2003) *Journal of Bacteriology* **185**, 2251-2258.
4. D'Aquino, J. A. & Ringe, D. (2003) *Journal of Bacteriology* **185**, 4081-4086.
5. Spiering, M. M., Ringe, D., Murphy, J. R. & Marletta, M. A. (2003) *Proceedings of the National Academy of Sciences of the United States of America* **100**, 3808-3813.
6. Tao, X. & Murphy, J. R. (1992) *Journal of Biological Chemistry* **267**, 21761-21764.
7. Wang, Z. O., Schmitt, M. P. & Holmes, R. K. (1994) *Infection and Immunity* **62**, 1600-1608.
8. Tao, X., Zeng, H. Y. & Murphy, J. R. (1995) *Proceedings of the National Academy of Sciences of the United States of America* **92**, 6803-6807.

LIST OF PARTICIPANTS

LastName	FirstName	ORGANIZATION	COUNTRY
Abián Franco	Olga	BIFI, Univ. de Zaragoza	Spain
Alonso Buj	José L.	Dep. Física Teórica y BIFI, Universidad de Zaragoza	Spain
Amzel	Mario	The Johns Hopkins University, School of Medicine	USA
Antonova	Olga	Molecular and Radiation Biophysics Division, Petersburg Nuclear Physics Institute	Russia
Ares	Saúl	Max-Planck-Institut für Physik komplexer Systeme, Dresden	Germany
Arias Moreno	Xabier	Dep Bioquímica y Biología Molecular y Celular y BIFI, Universidad de Zaragoza	Spain
Ariza García	Luis Fernando	Dep. Química-Física, Universidad de Granada	Spain
Ayuso	Sara	Dep. Bioquímica y Biología Molecular y Celular y BIFI, Universidad de Zaragoza	Spain
Basdevant	Nathalie	Département de Physique et Modelisation	France
Bastolla	Ugo	Centro de Biologia Molecular Severo Ochoa, Madrid	Spain
Beke	Tamás	Depart Organic Chemistry, Eotvos Univ.	Hungary
Bernadó	Pau	Par Cientific de Barcelona, UB	Spain
Bes Fustero	María Teresa	Dep. Bioquímica y Biología Molecular y BIFI, Univ. Zaragoza	Spain
Bruscolini	Pierpaolo	BIFI, Universidad de Zaragoza	Spain
Bueno	Marta	Dep Bioquímica y Biología Molecular y Celular y BIFI, Univ. de Zaragoza	Spain
Calvo	Iván	Dep Física Teórica y BIFI, Univ. de Zaragoza	Spain
Campodominico	Paola	Universitat Jaume I, Castellón	Spain
Campos Plasencia	Isabel	BIFI, Universidad de Zaragoza	Spain
Casetti	Lapo	Dipartimento di Fisica and CSDC, Università di Firenze	Italy
Cavasotto	Claudio N.	Molsoft	USA
Clemente Gallardo	Jesús	BIFI, Univ. de Zaragoza	Spain
Clote	Peter	Biology Department, Boston College	USA
Cocco	Simona	Labratoire de Physique Statistique de L'ENS	France
Cotallo Abán	María	BIFI, Universidad de Zaragoza	Spain
Cremades	Nunilo	Dep Bioquímica y Biología Molecular y Celular y BIFI, Universidad de Zaragoza	Spain
Cruz	Andrés	Dep. de Física Teórica y BIFI, Universidad de Zaragoza	Spain
D' Aquino-Ruiz	Alejandro	MS 029 Brandeis University	USA
Dannenberg	Joseph	City University of New York -Hunter College	USA
De Fabritis	Gianni	Chemistry Department, University College London	UK
Dobnikar	Jure	University Graz	Germany
Dumas	Philippe	IBMC-CNRS Strasbourg	France
Echenique	Pablo	Dep Física Teórica y BIFI, Universidad de Zaragoza	Spain
Estrada	Jorge	Dep Bioquímica y Biología Molecular y Celular y BIFI, Universidad de Zaragoza	Spain
Faisca	Patricia	Centro de Fisica Teorica e Computacional	Portugal

Falceto	Fernando	Dep Física Teórica y BIFI, Univ. de Zaragoza	Spain
Falo Forniés	Fernando	Dep. Física de la Materia Condensada y BIFI, Univ. Zaragoza	Spain
Fernandez Alvarez-Estrada	Ramón	Dep Física Teórica, Universidad Complutense, Madrid	Spain
Fillat	Maria F.	Dep Bioquímica y Biología Molecular y Celular y BIFI, Univ. de Zaragoza	Spain
Floría Peralta	Luis Mario	Dep Física de la Materia Condensada y BIFI, Univ. de zaragoza	Spain
Frago	Susana	Departamento de Bioquímica y Biología Molecular y Celular y BIFI, Univ. de Zaragoza	Spain
Freire	Ernesto	Johns Hopkins Univ. Dep. of Biology	USA
Ganoth	Assaf	Tel Aviv University	Israel
García Esteve	José V.	Dep Física Teórica y BIFI, Univ. de Zaragoza	Spain
Geroult	Sebastien	Institut of Structural Molecular Biology	U.K
Ghozzi	Stephane	Laboratoire de Physique Statistique de l\'ENS	France
Godoy Ruiz	Raquel	Dep. Química-Física, Univ. de Granada	Spain
Gómez Gardeñes	Jesús	Dep de Física de la Materia Condensada y BIFI, Univ. de Zaragoza	Spain
González Navarrete	Patricio Andres	Univ. Jaume I de Castellón	Spain
Goñi Rasia	Guillermina M.	Dep Bioquímica y Biología Molecular y Celular y BIFI, Univ. de Zaragoza	Spain
Gordillo Guerrero	Antonio	Univ. de Extremadura	Spain
Gracia	Luis	Weill Medical College of Cornell University	USA
Gruziel	Magdalena	Warsaw University, Department of Biophysics	USA
Han	Rongsheng	Centro de Biología Molecular "Severo Ochoa", Madrid	Spain
Huertas Gambín	Oscar	Dep Físicoquímica Univ. de Barcelona	Spain
Ibarra Molero	Beatriz	Dep. Química-Física, Univ.d de Granada	Spain
Jang	Joonkyung	Pusan National University	Korea
Janson	Natalia	Loughborough University, UK	UK
Jaramillo	Alfonso	Laboratoire de Biochimie	France
Jiménez San Juan	Sergio	Dipartimento di Fisica. Univ di Roma I, La Sapienza.	Italy
Junier	Ivan	Univ. de Barcelona	Spain
Karplus	Martin	Dep. Chemistry Biology, Harvard Univ.	USA
KUMAR	Sanjay	Department of Physics, Banaras Hindu University	India
Leibler	Stanislas	The Rockefeller University	USA
López Cordón	Antonio Javier	Dep. Química-Física, Univ. de Granada	Spain
López Gomollón	Sara	Dep. Bioquímica y Biología Molecular y BIFI, Univ. Zaragoza	Spain
López Ruiz	Ricardo	Dep Informática e Ingeniería de Sistemas y BIFI, Univ. de Zaragoza	Spain
Louzoun	Yoram	Bar Ilan University	Israel
Luque	F. Javier	Univ. de Barcelona	Spain
Mahesh	Kulharia	MPI for Molecular Physiology	Germany
Marinari	Enzo	Dipartimento di Fisica, Università di Roma La Sapienza	Italy
Martín	José Manuel	Dep. de Química-Física, Fac. de Ciencias, Univ. de Granada	Spain
Martín Luna	Beatriz	Dep. Bioquímica y Biología Molecular y BIFI, Univ. Zaragoza	Spain
Martínez Júlvez	Marta María	Dep. Bioquímica y Biología Molecular y Celular y BIFI, Univ. de Zaragoza	Spain

Matijssen	Berry	Trinity College Dublin	Ireland
Medina	Milagros	Dep de Bioquímica y Biología Molecular y Celular y BIFI, Univ. de Zaragoza	Spain
Morel	Bertrand	Dep. Química-Física, Fac. de Ciencias, Univ. de Granada	Spain
Moreno Vega	Yamir	BIFI, Univ. de Zaragoza	Spain
Morreale	Antonio	Unidad de Bioinformática, Centro de Biología Molecular Severo Ochoa	Spain
Nahum Alves	Cláudio	Univ. Jaume I de Castellón	Spain
Palencia	Andrés	Dep. Química-Física, Fac. de Ciencias, Univ. de Granada	Spain
Pálfi	Villö	Depart Organic Chemistry, Eotvos University	Hungary
Parody Morreale	Antonio	Dep. Química-Física, Univ. de Granada	Spain
Peleato Sánchez	María Luisa	Dep. Bioquímica y Biología Molecular y BIFI, Univ. Zaragoza	Spain
Pellicer Espuña	Silvia	Dep. Bioquímica y Biología Molecular y BIFI, Univ. Zaragoza	Spain
Perczel	András	EOTVOS University	Hungary
Peregrina	José Ramón	Dep Bioquímica y Biología Molecular y Celular y BIFI, Univ. de Zaragoza	Spain
Polo	Victor	Univ. Jaume I de Castellón	Spain
Prada	Diego	Dep. Física de la Materia Condensada y BIFI, Uiv. Zaragoza	Spain
Ramirez Ortiz	Angel	Unidad de Bioinformática, Centro de Biología Molecular Severo Ochoa	Spain
Recacha Castro	Rosario	Department Biochemistry, Trinity College, Dublín	Ireland
Ritort	Félix	Dep. de Física Fundamental, UB	Spain
Rodríguez Larrea	David	Dep. Química-Física, Univ. de Granada	Spain
Rossi	Barbara	Department of Physics, University of Trento	Italy
Ruiz Lorenzo	Juan J.	Univ. de Extremadura	Spain
Samsonov	Sergey	Faculty of Physical Science	Russia
Sánchez	Angel	GISC/Matemáticas, Universidad Carlos III de Madrid	Spain
Sánchez Ruiz	José Manuel	Dep. Química-Física	Spain
Sancho	Javier	Dep Bioquímica y Biología Molecular y Celular y BIFI, Univ. de Zaragoza	Spain
Serrano	Ana	Dep Bioquímica y Biología Molecular y Celular y BIFI, Univ. de Zaragoza	Spain
Setny	Piotr	Warsaw Uniwersity, Department of Biophysics	Poland
Sevilla Miguel	Emma	Dep.Bioquímica y Biología Molecular y BIFI, Univ. Zaragoza	Spain
Singh	Navin	Department of Physics, Banaras Hindu University	India
Sukovataya	Irina	Dept.of Biophysics, Krasnoyarsk State Univ.	Russia
Sussman	Fredy	Univ. de Santiago de Compostela	Spain
Tamm	Mikhail	Phys. Dept., Moscow State University	Russia
Tarancón Lafita	Alfonso	Dep Física Teórica y BIFI, Univ. de Zaragoza	Spain
Tortosa	Pablo	Ecole Polytechnique, Paris	France
Tramontano	Anna	Dipartimento di Scienze Biochimiche	Italy
Turdikulova	Shakhlokhon	National University of Uzbekistan	Uzbekistan
Urrutia	Mariela	Instituto Leloir	Argentina
Velásquez Campoy	Adrián	BIFI, Univ. de Zaragoza	Spain
Verrocchio	Paolo	Department of Physics, University of Trento	Italy

Vidal	David	Parc Cientific de Barcelona, UB	Spain
Viliani	Gabriele	Department of Physics, University of Trento	Italy
Villa Bernáldez	Isaac	Dep Química Física. Univ. de Sevilla	Spain
Waksman	Gabriel	Institute of Structural Molecular Biology at UCL	UK
Wall	Michael E.	Los Alamos National Laboratory	USA
Westhof	Eric	Institut de Biologie Moléculaire et Cellulaire, Strasbourg	France
Zaman	Muhammad	Computational & Systems Biology	USA

AUTHOR INDEX

A

Alonso, J. L., 108
Amzel, L. M., 1

B

Basdevant, N., 192
Bes, M. T., 129
Borgis, D., 192
Buscolini, P., 135

C

Calvo, I., 108
Cavasotto, C. N., 34
Cavelier, G., 1
Cocco, S., 50
Cotallo-Abán, M., 135

D

Dannenberg, J. J., 102
D'Aquino, J. A., 196

E

Echenique, P., 108

F

Falo, F., 135
Fillat, M. F., 129
Floría, L. M., 150
Frago, S., 180

G

Giorgetti, A., 64
Gómez-Gardeñes, J., 150
Gracia, L., 176

H

Ha-Duong, T., 192
Hernández, J. A., 129

J

Jaramillo, A., 96
Junier, I., 70

L

López-Gomollón, S., 129

M

Martín, B., 129
Martínez-Júlvez, M., 180
Mazo, J. J., 135
Medina, M., 180
Monasson, R., 50
Moreno, Y., 150

N

Nordlund, E., 185

P

Peleato, M. L., 129
Pellicer, S., 129
Platonova, N. A., 185
Prada-Gracia, D., 135
Puchkova, L. V., 185

R

Raimondo, D., 64
Ringe, D., 196
Ritort, F., 70

S

Samsonov, S. A., 185
Sanchez-Ruiz, J. M., 123
Sancho, J., 135
Schubert, L., 180
Sevilla, E., 129
Skvortsov, A. N., 185
Speidel, J. A., 176

T

Tortosa, P., 96
Tramontano, A., 64
Tsymbalenko, N. V., 185

V

Velázquez-Campoy, A., 162

W

Wall, M. E., 16
Weinstein, H., 176

Z

Zaman, M. H., 117